Global Environmental Change: Challenges to Science and Society in Southeastern Europe

Vesselin Alexandrov · Martin Felix Gajdusek ·
C. Gregory Knight · Antoaneta Yotova
Editors

Global Environmental Change: Challenges to Science and Society in Southeastern Europe

Selected Papers presented in the International Conference held 19–21 May 2008 in Sofia Bulgaria

Springer
SAFE
ZSI ASO LJUBLJANA SOFIA AUSTRIAN SCIENCE AND RESEARCH LIAISON OFFICE
BM.W_F[a] Austrian Federal Ministry of Science and Research

Editors
Dr. Vesselin Alexandrov
National Institute of
Meteorology & Hydrology
Tzarigradsko shose Blvd. 66
1784 Sofia
Bulgaria
Vesselin.Alexandrov@meteo.bg

Dr. Martin Felix Gajdusek
Austrian Science and Research
Liaison Office Sofia
ASO Sofia
ul. Moskovska 5
1000 Sofia
Bulgaria

Dr. C. Gregory Knight
Pennsylvania State University
Department of Geography
302 Walker Building
University Park PA 16802
USA

Antoaneta Yotova
National Institute of
Meteorology & Hydrology
Tzarigradsko shose Blvd. 66
1784 Sofia
Bulgaria

ISBN 978-94-007-9459-7 ISBN 978-90-481-8695-2 (eBook)
DOI 10.1007/978-90-481-8695-2
Springer Dordrecht Heidelberg London New York

Printed on acid-free paper

Springer is part of Springer Science+Business Media (www.springer.com)

Preface

This book is the outcome of the conference "Global Environmental Change: Challenges to Science and Society in Southeastern Europe" organized by the Scientific Coordination Center for Global Change (SCCGC) at the Bulgarian Academy of Sciences jointly with the Austrian Science and Research Liaison Offices Ljubljana and Sofia (ASO). The event was supported by the Federal Ministry of Science and Research of the Republic of Austria in the framework of its SEE science cooperation initiative. Sponsorship was also provided from the Knight-Staneva Foundation for Sustainability and Future Environments (USA) and the EC FP6 ADAGIO project (www.adagio-eu.org).

The Conference celebrated the anniversary of the founding of the SCCGC in 1997 as National Coordination Center for Global Change (NCCGC). The idea for such a center evolved at the June 1997 Workshop "Global Change and Bulgaria" held in the American University in Bulgaria, Blagoevgrad, sponsored by the US National Science Foundation via the Center for Integrated Regional Assessment (CIRA) at the Pennsylvania State University. In examining the competencies Bulgarian scientists would bring to the study of climate change and its impacts, that workshop identified a number of research priorities and future strategies to be pursued. The workshop resulted in two publications, Глобалните промени и България[1] and *Global Change and Bulgaria.*[2]

The most important future strategy from the 1997 Workshop was the proposal to the leadership of the Bulgarian Academy of Sciences for formation of the NCCGC, with the recommendation that it be headed by Academician Dimitar Mishev, director of the Academy's Solar-Terrestrial Influences Laboratory, and membership from a variety of institutes and other organizations. In July 1997, the NCCGC came into being and operated under that title until the death of its president, Academician Mishev, in 2003.

---

[1] 1999. Sofia: National Coordination Center for Global Change, Bulgarian Academy of Sciences, 370 pp. (Todor Hristov, C. Gregory Knight, Dimitar Mishev, Marieta P. Staneva, editors; in Bulgarian). ISBN 954-90485-1-9.

[2] 2000. Sofia: National Coordination Center for Global Change, Bulgarian Academy of Sciences, 350 + viii pp. (Marieta P. Staneva, C. Gregory Knight, Todor N. Hristov, Dimitar Mishev, editors). ISBN 954-90485-2-7.

During that interim, one of the research priorities was initiated, namely a study of the 1982–1994 Bulgarian drought as an analog of future climate change, realized as a project coordinated in Bulgaria by Professor Ivan Raev of the Forest Research Institute and in the US by C. Gregory Knight of the Pennsylvania State University, with funding from the US National Science Foundation via CIRA. That project also resulted in two books, Засушаването в България: съвременен аналог за климатични промени[3] and *Drought in Bulgaria: A Contemporary Analog of Climate Change.*[4]

The NCCGC played an important role in the creation of the Industrial Transformation Science Plan of the International Human Dimensions Programme in 1998–1999. The Centre also co-sponsored a workshop on Integrated Regional Assessment of Climate Change held in Budapest in 1999 and hosted the Sofia workshop on Human Dimensions of Global Change in 2000.

Subsequent to the death of Academician Mishev and by decision of the Board of the Bulgarian Academy of Sciences, NCCGC became the SCCGC in 2003 with Professor Ivan Raev as president. The SCCGC continues until the present as a consultative and advisory body on global change issues in Bulgaria with the presidency having been handed from Professor Raev to Associate Professor Vesselin Alexandrov, director of the Meteorology Department of NIMH, in May 2006.

In 2007, in recognition of the 10th anniversary of the workshop that had led to the creation of the SCCGC, plans were begun to organize a conference on global environmental change issues in the region of Southeastern Europe in co-operation with the ASO. The conference was held in Park Hotel Moskva, Sofia, on 19–21 May 2008 with near 120 participants, including 36 from outside Bulgaria (Albania, Austria, Bosnia and Herzegovina, Croatia, Cyprus, Czech Republic, Finland, FYR of Macedonia, Greece, Latvia, Moldova, Montenegro, Serbia, Slovenia, Romania and USA), which reflected the interest in environmental change issues throughout Southeastern Europe and beyond. There were four plenary keynotes, eight thematic parallel and poster sessions during the conference.[5]

This book is a collection of papers presented in the conference as selected from presentations and respectively revised for publication. The editors and authors of contributions hope that the volume will be a way marker on the road of basic and applied research on climate change and other environmental issues as well as related

---

[3]2003. Sofia: Bulgarian Academy of Sciences Press, 284 pp. (Ivan Raev, C. Gregory Knight, Marieta P. Staneva, editors). ISBN 954-90896-1-4.

[4]2004. Aldershot, UK: Ashgate Studies in Environmental Policy and Practice. 336 + xvi pp. (C. Gregory Knight, Ivan Raev, Marieta P. Staneva, editors). ISBN 978-0754642152.

[5]For further information refer to http://global-change.meteo.bg/conference_en.htm or refer to http://www.aso.zsi.at/bg/veranstaltung/2601.html (last accessed on September 4, 2009).

impacts in Southeastern Europe, and that it will also introduce some of the work done in this region to the global change community.

Vesselin Alexandrov
Martin F. Gajdusek
C. Gregory Knight
Antoaneta Yotova

Contents

Contributors

Vesselin Alexandrov National Institute of Meteorology and Hydrology, Sofia, Bulgaria, vesselin.alexandrov@meteo.bg

Samar Altaher Central Laboratory for Agricultural Climate (CLAC), Giza, Egypt, sattaher2001@yahoo.com

Dimos P. Anastasiou Institute of Environmental Research and Sustainable Development (IESRD-NOA), Athens, Greece, dimos_anastasiou@yahoo.com

Blanka Bartók Faculty of Geography, University of Babeş-Bolyai, Cluj-Napoca, Romania, bartokblanka@yahoo.com

Viorel Blujdea JRC – Institute for Environment and Sustainability, Ispra, Italy, viorel_blujdea@yahoo.com

Lilia Bocheva National Institute of Meteorology and Hydrology, Sofia, Bulgaria, Lilia.Bocheva@meteo.bg

Lučka Kajfež Bogataj University of Ljubljana, Ljubljana, Slovenia, lucka.kajfez.bogataj@bf.uni-lj.si

Eglantina Bruci Climate Change Programme, Ministry of Environment, Forests and Water Administration, Tirana, Albania, eglantina.bruci@undp.org

George Cojocaru TIAMASG Foundation, Bucharest, Romania, gco@icpa.ro

Nataliya Dimitrova-Popova Rio Conventions Project, United Nations Development Programme, Sofia, Bulgaria, Natalia@rioconventions.org

Josef Eitzinger University of Natural Resources and Applied Life Sciences, Vienna, Austria, josef.eitzinger@boku.ac.at

Ilian Gospodinov National Institute of Meteorology and Hydrology, Sofia, Bulgaria, ilian.gospodinov@meteo.bg

Gregor Gregorič Environmental Agency of Slovenia, Ljubljana, Slovenia, gregor.gregoric@gov.si

Tomas Halenka Department of Meteorology and Environment Protection, Charles University, Prague, Czech Republic, tomas.halenka@mff.cuni.cz

Juha Hatakka Finnish Meteorological Institute, Helsinki, Finland, Juha.Hatakka@fmi.fi

Doru Leonard Irimie Ministry of Agriculture, Forests and Rural Development, Bucharest, Romania, doru.irimie@madr.ro

C. Gregory Knight Department of Geography, Pennsylvania State University, University Park, PA, USA, cgk@psu.edu

Judith Köberl Joanneum Research, Institute of Technology and Regional Policy, Graz, Austria, judith.koeberl@joanneum.at

Stanimir Kostadinov Faculty of Forestry, Belgrade University, Belgrade, Serbia, kost@eunet.rs

Gerhard Kubu University of Natural Resources and Applied Life Sciences, Vienna, Austria, gerhard.kubu@boku.ac.at

Branislava Lalic Centre for Meteorology and Environmental Predictions (CMEP), University of Novi Sad, Novi Sad, Serbia, branka@polj.ns.ac.yu

Jacek Leśny Agrometeorology Department, August Cieszkowski Agriculture University of Poznan (ACAUP), Poznan, Poland, jlesny@up.poznan.pl

Željko Majstorović Federal Hydrometeorological Institute, Sarajevo, Bosnia and Herzegovina, zemajstorovic@yahoo.com

Tania Marinova National Institute of Meteorology and Hydrology, Sofia, Bulgaria, tania.marinova@meteo.bg

Slađan Maslać Western Balkans Environmental Programme, United Nation Development Programme Montenegro, Podgorica, Montenegro, okolina@inneco.net

Mahmoud Medany Central Laboratory for Agricultural Climate (CLAC), Giza, Egypt, rumedany@yahoo.com

Dragutin T. Mihailovic Faculty of Agriculture, Centre for Meteorology and Environmental Predictions (CMEP), University of Novi Sad, Novi Sad, Serbia, guto@polj.ns.ac.yu

Plamen Mishev University of National and World Economy, Sofia, Bulgaria, m.mochurova@iki.bas.bg

Milkana Mochurova Institute of Economics, Bulgarian Academy of Sciences, Sofia, Bulgaria, m.mochurova@iki.bas.bg

Suzana Alcinova Monevska Hydrometeorological Institute, Skopje, Republic of Macedonia, smonevska@meteo.gov.mk

Liri Muçaj Institute for Energy, Water and Environment, Tirana, Albania, liri_mucaj@yahoo.com

Vangjel Mustaqi Institute for Energy, Water and Environment, Tirana, Albania, vmustaqi@yahoo.com

Natalia Nemeshko State Hydrological Institute (SHI), St. Petersburg, Russia, natlem@mail.ru

Dragana Nikić Faculty of Medicine, University of Niš, Niš, Serbia, mani@junis.ni.ac.rs

Michael Nikolaev Agrophysical Research Institute (ARI), St. Petersburg, Russia, clenrusa@mail.ru

Maja Nikolić Faculty of Medicine, University of Niš, Niš, Serbia, mani@junis.ni.ac.yu

Janusz Olejnik Agrometeorology Department, August Cieszkowski Agriculture University of Poznan (ACAUP), Poznan, Poland, olejnikj@au.poznan.pl

Jussi Paatero Finnish Meteorological Institute, Helsinki, Finland, jussi.paatero@fmi.fi

Ciprian Pahonţu National Forest Administration ROMSILVA, Bucharest, Romania, ciprian.pahontu@maa.ro

Stoytcho Panchev Solar Terrestrial Influences Institute, Sofia, Bulgaria, spanchev@phys.uni-sofia.bg

Franz Prettenthaler Joanneum Research, Institute of Technology and Regional Policy, Graz, Austria, franz.prettenthaler@joanneum.at

Pece Ristevski Hydrometeorological Institute, Skopje, Republic of Macedonia, smonevska@meteo.gov.mk

Daniela Semeradova Mendel University of Agriculture and Forestry in Brno (MZLU), Brno, Czech Republic, daniela.semeradova@mendelu.cz

Petio Simeonov National Institute of Meteorology and Hydrology, Sofia, Bulgaria, petio.simeonov@meteo.bg

Catalin Simota TIAMASG Foundation, Bucharest, Romania, csimota@cpa.ro

Igor G. Sirodoev Institute of Ecology and Geography, Moldavian Academy of Sciences, Chisinau, Moldova, ingvarri@gmail.com

Tatiana Spassova National Institute of Meteorology and Hydrology, Sofia, Bulgaria, tatiana.spassova@meteo.bg

Marieta P. Staneva Department of Geography, Pennsylvania State University, University Park, PA, USA, mps5@psu.edu

Andreja Sušnik Environmental Agency of Slovenia, Ljubljana, Slovenia, andreja.susnik@gov.si

Sabina Thaler University of Natural Resources and Applied Life Sciences, Vienna, Austria, sabina.thaler@boku.ac.at

Miroslav Trnka Mendel University of Agriculture and Forestry in Brno (MZLU), Brno, Czech Republic, mirek_trnka@yahoo.com

Angel Utset Agrarian Technological Institute of Castilla and Leon (ITACYL), Valladolid, Spain, autset@clnearth.com

Blagorodka Veleva National Institute of Meteorology and Hydrology, Sofia, Bulgaria, BLAGORODKA.VELEVA@meteo.bg

Domenico Ventrella Instituto Sperimentale Agronomico (CRA-ISA), Bari, Italy, domenico.ventrella@entecra.it

Stewart Williams Western Balkans Environmental Programme, United Nation Development Programme Montenegro, Podgorica, Montenegro, stewart.williams@undp.org

Zdenek Zalud Mendel University of Agriculture and Forestry in Brno (MZLU), Brno, Czech Republic, zdenek.zalud@mendelu.cz

Dženan Zulum Federal Hydrometeorological Institute, Sarajevo, Bosnia and Herzegovina, dzenanz@fhmzbih.gov.ba

Abbreviations

CEU	Central European University
EA	Environmental Assessment
EIA	Environmental Impact Assessment
EU	European Union
GE	Global Environment
GEF	Global Environmental Facility
GIS	Geographic Information System
MoEW	Ministry of Environment and Water
MRDPW	Ministry for Regional Development and Public Works
NCSA	National Capacity Self Assessment
NGO	Non-Governmental Organization
NOPRD	National Operational Program for Regional Development
NSFRD	National Strategy for Regional Development
NTTA	Natura 2000 Assessment
OPs	Operational Programs
RCP	Rio Conventions Project
RDA	Regional Development Act
RDPs	Regional Development Plans
SA	Sustainability Appraisal
SD	Sustainable Development
SEA	Strategic Environmental Assessment
SEE	South Eastern Europe
SU	Sofia University
TP	Training Program
UD&SP	Urban Development and Spatial Planning
UNCBD	United Nations Convention on Biological Diversity
UNCCD	United Nations Convention to Combat Desertification
UNDP	United Nations Development Programme
UNEP	United Nations Environment Programme
UNFCCC	United Nations Framework Convention on Climate Change

Introduction

Vesselin Alexandrov, Martin F. Gajdusek, C. Gregory Knight, and Antoaneta Yotova

Global change is the term used to encompass a multitude of environmental and ecological changes of different scale that have been noticed, measured and studied on the Earth. Global environmental change encompasses the study of issues such as climate change, biodiversity and species extinction, land use change, changes in the carbon and water cycles, air quality, etc. Global environmental change is as old as the Earth because physical, chemical and biological processes have been shaping and reshaping the planet's environment since its infancy 4.5 billion years ago. In recent time, however, humankind has become one of the major driving forces of environmental change on our planet by diverse activities resulting in climate change, loss of biodiversity, pollution, desertification and other consequences. In turn, both individuals and societies are also experiencing the impact of the changes in their natural environment upon their own social, economic and political situations.

In the region of South-Eastern Europe, changing weather patterns are creating droughts, floods, and forest fires. Severe droughts affected most socio-economic sectors in this region in the twentieth century. Since 2002, floods in Europe have caused a number of deaths, the displacement of about half a million people, and cost at least 25 billion Euro as measured by insured economic losses. In addition to wet weather conditions, poor river basin management contributes to flooding in South-Eastern Europe. Furthermore, water leakages from distribution networks are frequent in many South-Eastern European countries. It is common that more than one third of the supplied water is lost before delivery.

All countries in South-Eastern Europe are parties to the UN Framework Convention on Climate Change of 1992. Between 2004 and 2007, countries such as Albania, Bulgaria, Croatia, Former Yugoslavian Republic (FYR) of Macedonia and Montenegro ratified the Kyoto Protocol. Although the absolute and per capita contribution of the region to global greenhouse gas emissions is relatively small, there are significant opportunities for emission reductions. Countries can reduce their current high energy consumption and energy intensity, lower their reliance on coal for energy production, increase energy efficiency of the domestic sector, and invest in renewable energy resources such as solar, biomass, hydro, and wind.

Biodiversity is under serious threat in South-Eastern Europe, particularly in farmland, mountain regions and coastal zones. The loss of biodiversity happens primarily because of human activities leading to land use changes – including urban sprawl, infrastructure development, intensification and/or abandonment of the agricultural sector, as well as to acidification, eutrophication, desertification, overexploitation of natural resources, and climate change.

The countries of South-Eastern Europe have made substantial progress in developing legal and policy reforms in the area of environmental protection, most of it linked to EU accession or membership. However, the establishment of appropriate institutional infrastructures often remains a challenge. Countries of the region still face lack of sufficient resources and adequate administrative capacity for adoption and effective implementation of the multilateral environmental agreements. The quality of South-Eastern Europe's nature and environment is high but in order to preserve this status, the concept of sustainable development must be practically integrated into the mainstream economic and social development policies.

In the last decade, the environmental changes and awareness of their impacts on society, economy, and local and regional development have become important research topics in South-Eastern Europe. Global environmental change and its implications for the future development in the region are nowadays hot issues for politicians, and this impetus accelerates the introduction of specific adaptation and mitigation measures. Climate change is one of the most important public and society driven research topics but also one of the very first research issues requiring a real interdisciplinary, and therefore challenging, approach. During recent years, global change research and participation of researchers from South-Eastern Europe in the global change community have grown. Research networks in related fields emerged recently, driven also by diverse funding schemes, e.g. programmes of the European Union. The explicit societal demand challenges researchers and their science communication competencies on the research topics concerned. Thus, it became necessary to convene a meeting to assess the recent knowledge on global environmental change and related impacts in South-Eastern European countries. The international conference "Global Environmental Change: Challenges to Science and Society in South-Eastern Europe", held in May 2008 in Sofia, Bulgaria, aimed to respond this need. The particular interest on the region of South-Eastern Europe is because recently and in the future, this region will become increasingly sensitive to global changes. Beyond the work of scientists and actions of governments, civil society has a major stake in such changes, so the necessity for information exchange as basis for actions to address resulting stress as well as to exploit new opportunities is important task.

As main objectives of the above conference, different aspects of the environmental changes and their impacts are discussed in the next parts of this book. It starts with the four key-note lectures presented during the conference. In these chapters of the book, attention is paid to the possibility for numerical experiments with climatic models and to discovery of local and regional "signature" of the global climatic changes. The importance of a focused Balkan-region analysis of future climate change and climate change impacts on a more detailed regional basis than undertaken before is considered. Climate change scenarios based on high resolution climate models for Europe and the possible impacts of climate change due to changes in extreme weather and climate events for different sectors (agriculture, water resources, human health, built environment and infrastructure) are discussed. The last key-note chapter traces the history of the establishment of the Scientific Coordination Center for Global Change in Bulgaria aiming to ensure research projects, studies, publications and educational initiatives in the realm of global change and goal for sustainable development. Research results and lessons learned are viewed as road markers to assist decision makers and the public particularly in the face of climate change and planning for sustainability.

Global Change and Climate Change in South-Eastern Europe

In this part of the book, the chapters represent selected research on environmental changes at global, regional and local scales in the past, present and future including assessments of observations, indices and indicators, models, scenarios, trends, etc. In the first chapter, some issues which are important from the point of view of the current climate modelling attempts are considered, namely: (i) overview of the results achieved until the present time; (ii) discussion of the term predictability beyond complexity; (iii) consideration of an approach in analysis of the energy balance equation for the environmental interface and (iv) "chaotic" response of the environmental interface due to forcing by visible and infrared radiation. The aim of the second chapter is to estimate regional changes of different type of clouds associated with climate change in the south-eastern part of Europe. A statistical method of downscaling is applied to calculate how cloudiness change in parallel with the increase of hemispherical temperature with 0.5°K; the regression of local variables against the hemispherical mean temperature in the recent monotonously warming period of 1973–1996 is analyzed by the method of instrumental variables. The next chapter comprises recent achievements and updated research results on projected climate change in FYR of Macedonia in the twenty-first century, related to previously elaborated National Communications on Climate Change as an obligation under the UN Framework Convention on Climate Change. Then, authors from Bosnia and Herzegovina present the effects of climate changes, such as increase of average temperature about 0.7°C in the past 100 years and the annual sum of rainfalls with no severe changes, finding the existence of very warm and very cold short periods, and also their consecutive fast changes, over the country's territory. In the last chapter of this part, the historical archive of synoptic maps and NCEP/NCAR reanalysis data files are used for analysis and classification of the synoptic situations causing torrential precipitation over the territory of Bulgaria; the fields of air pressure and wind velocity are also considered. A negative trend in the annual and seasonal precipitation totals associated with an increase in the contribution of heavy rainfall events to the total precipitation is observed in the country as in some Mediterranean countries.

Environmental Impacts

The impacts of global, regional and local environmental changes on natural systems, such as water resources, forests, and ecosystems, are subject of this part of the book. The first two chapters present Joanneum research in the field of climate change – scientific and network activities in Austria and South-Eastern Europe. In the next chapter, a specific targeted research project CECILIA (Central and Eastern Europe Climate Change Impact and Vulnerability Assessment) is presented. The project's main goal is to provide climate change impacts and vulnerability assessment in targeted areas of Central and Eastern Europe. To assess the impact of climate change at the regional to local scale, very high resolution Regional Climate Models are run locally for targeted areas in order to capture the effects of the complex terrain of the region. Changes in weather patterns and extreme events are addressed within the project as they affect important economic sectors and welfare of individual countries in the region. The selected applications of the CECILIA outputs are supposed toward water resources management, agriculture, forestry, air quality and health. The next chapter discusses results of the regional studies and feedback gathered from experts and farmers which show in general that drought and heat are the main factors having impact on agricultural vulnerability not only in the Mediterranean region, but also in the central and eastern European regions. Another important aspect is that the increasing risk of pest and diseases may play a more important role for agricultural vulnerability than assumed before, however, till now this field is only rarely investigated in Europe. Review of climate change projections based on a range of emission scenarios extending up to the end of the twenty-first century is made in the next chapter. The study area covers South-Eastern European countries, such as Croatia, Bosnia and Herzegovina, Serbia, Montenegro, Albania and the FYR of Macedonia. The scenarios exhibit an anticipated increase in the annual air temperature higher than 5°C with very slight differences over the region. As result of the reduction in annual total precipitation rate, the study area could experience a general decrease in runoff. The demand for water could increase, especially in the summer. It could cause more keen problems, particularly for energy production from hydropower in countries like Albania that highly depend on hydropower. The state of climate changes in Serbia and their consequences: desertification, soil erosion, torrential floods, as well as the significance and effects of forests on their mitigation, are subjects of the next chapter. Erosion, erosion sediment and torrential floods cause numerous damages among which the damage brought to the environment is increasingly important. It is manifested as landscape degradation, mechanical and chemical pollution of water in watercourses and storages. With present climate changes, which have an unfavourable impact on the development of vegetation, the risk of intensified erosion processes is increased.

Impacts on Socio-Economic Activities

In this part of the book, the impacts of environmental changes on society and economy, for example impacts on human health, agricultural production, other industrial and economic sectors, are illustrated by selected chapters. Chapter 15 outlines the conceptual framework of the different types of climate change impacts and justifies the necessity of economic impact studies, especially on a local level. Results of a number of impacts studies at global, European and country level have been examined, not to give single values of damage or impact of climate change, but to explore plausible ranges of impacts. The chapter concludes with the expected results of a research project – CLAVIER (Climate Change and Viability: Impacts on Central and Easter Europe) aiming at filling the research gap concerning local level studies related especially to economic impact and vulnerability issues. The next chapter is devoted to the heavy rains in Albania which are amongst the greatest problems nowadays, especially on the low and coastal land. These zones very often are suffering from flooding caused not only by heavy rain but also by poor management of the urban infrastructure. Human factors are determinant in the frequent inundation of the coastal zones, as listed in the conclusion of the chapter. To analyze in details heavy rain and its social impact, the region around Lezha station as a representative for the coastal northern part of Albania is chosen. The threshold calculation for adverse weather phenomena identification and frequency distribution of meteorological variables is used as methodological tool. In Chapter 17 of this part, effects on the red blood cells in children exposed to air pollution are evaluated using the example of Nish in Serbia. The diagnosis of iron deficiency anemia is made using the pre-defined criteria. The air concentrations of black smoke, nitrogen dioxide, sulfur dioxide and lead in sediment matter are determined for the period from 1990 to 2000. The findings suggest that air pollution could have negative effects on red blood cells in children. Changes of water covered areas in Moldova during the transition period are analysed in the next chapter to reveal recent tendencies and find possible solutions of adaptation to climate change. "Spontaneous" concentration of reservoirs and ponds around big cities is found as a distinct feature of recent change. At the same time, the water-covered area diminished more drastically in initially water-scarce regions. As a result, adaptation measures must be territorially and typologically differentiated in order to achieve maximum effect and diminish social vulnerability of the population. Chapter 19 shows an observed trend which may indicate that the climate is becoming more maritime in northeastern and especially southeastern Europe. Namely, starting from the mid 1980s, there is a continuous decrease in the average lead-210 content of ground-level air both in Finland and Bulgaria, but the decreasing trend is much stronger in Bulgaria compared to Finland. However, the source areas of air masses are not the only factor to determine the lead-210 content of the air. The effect of the precipitation causing wet deposition of lead-210 has to be studied too, as the changes in the precipitation amount can influence the average residence time of lead-210 in the air.

From Global to Local and Vice Versa

The permanent direct and feedback links between global and smaller scale environmental changes and their impacts are discussed in this part of the book as illustrated by the selected chapters. Chapter 20 argues that in a changing climate, natural hazards are more frequent, and it would be reasonable to imply that social and economic risk is consequently increasing. Assessment of both – natural hazards and society vulnerability – belongs among the core work objectives of the Drought Management Centre for Southeastern Europe established in Slovenia. Historical assessment of drought occurrence and development of a drought monitoring system aim at establishment of regional products for estimation of climatological and actual natural hazards connected to drought. Some aspects of vulnerability have already been described for some of the countries in South-Eastern Europe. The next chapter states that one of the most important tasks of the signatory parties to the United Nations Convention to Combat Desertification (UNCCD) is to adopt National Action Programmes for combating desertification (NAP). The Romanian Government has adopted the first NAP in 2000. Following the previous experience with the limited implementation of this Programme, while confronting with the severe drought of 2007, it has been decided to proceed with its revision taking into account that (1) the political will and social impact of a revised NAP relates directly to the duration and intensity of the natural phenomenon (i.e., drought), and (2) bringing together a multitude of national authorities presumably concerned about drought and desertification incurs transaction costs which may easily exceed those required by the mere elaboration of NAP. Actions and measures proposed within this strategy may overlap with those provisioned by similar strategies (e.g., climate change mitigation/adaptation, water management, etc.), so that the critical issue of synergistic action comes again to the fore. The next chapter reports that the UNDP Country Offices in Albania, Bosnia and Herzegovina, Macedonia FYR, Montenegro, Serbia and the UN Administered Province of Kosovo have developed a regional demonstration programme around demand-driven projects in nine locations in the Western Balkans suffering from the legacy of polluting industries and requiring industrial renewal, environmental cleanup and new economic initiatives. The programme is to achieve improvement of the environmental situation and quality of life for citizens living in and around polluted areas through least-cost measures, improved local and national policy dialogue and supply of domestic professional services in the environmental management sector. While the main focus will be the physical works needed to mitigate the ecological problems, institutional strengthening and capacity building will be an important subject running throughout the programme. The last chapter presents the project "Integrating Global Environmental Issues into Bulgaria's Regional Development Process" initiated in 2006 by the Ministry of Regional Development and Public Works of Bulgaria and the United Nations Development Programme. The project's objective is to build capacities of respective ministries, district administrations and municipalities for mainstreaming global environmental issues into the formulation and implementation of regional and local development, as well as spatial planning policies. It is expected that by the end of the project, the beneficiaries shall have: adequate skills necessary for effective mainstreaming of global environmental issues into regional development policies; access to appropriate systems of training, mentoring, and learning in place to maintain a continuous skills upgrade of personnel; a comprehensive set of indicators that can be used to assess the impact of development and spatial planning at regional, district and municipal levels on the achievement of the UN Rio Conventions' objectives; access to data needed for reporting on progress in terms of impact on global environmental commitments; and models for update of regional development and spatial development plans and strategies.

We anticipate that the book chapters will be useful not only for experts and specialists in the field of global change and related impacts, but also for students and common people interested in environmental issues, especially in the region of South-Eastern Europe.

Part I
Keynotes

Chapter 1
Weather and Climate – Difficult Science Problems

Stoytcho Panchev and Tatiana Spassova

Abstract The classical physics with its fundamental principles and laws is in the basis of the contemporary weather and climate theories. The weather is defined as momentary state of the system Atmosphere–Land–Ocean (ALO). In this chapter, the development of the numerical method for deterministic weather forecasting is traced out and the main reasons leading to limited predictability are outlined. The climate is defined as a statistical ensemble of the ALO-system states for a long enough period of time τc ($\approx$30 years). The role of physics in the climate studies has two aspects:

1. Quantitative reconstruction of the past climate of our planet;
2. Development of theories and methods capable of reproducing the past climate events and of predicting future ones.

Special attention is paid to the possibility for numerical experiments with climatic models and to discovery of local and regional manifestation of the global climatic changes. The Bulgarian contribution to these current problems of physics and of the contemporary science in general, is also mentioned.

Keywords Climate · Numerical models · Weather

1.1 Introduction

About forty years ago, the great Russian scientist A. S. Monin (1969) published a book on physics and weather. Recently, he was a leading author of a review paper (Monin and Shishkov, 2000) on physics and climate. We share completely the ideas and concepts of this author concerning the crucial role of the physics in formulation and solution of the fundamental questions related to the weather and climate in cognitive and practical aspects. More precisely, we mean the classical physics with

S. Panchev (✉)
Solar Terrestrial Influences Institute, Sofia, Bulgaria
e-mail: spanchev@phys.uni-sofia.bg

V. Alexandrov et al. (eds.), *Global Environmental Change: Challenges to Science and Society in Southeastern Europe*, DOI 10.1007/978-90-481-8695-2_1,

its general principles and laws, applied to the multi-component system Atmosphere-Land-Ocean (ALO) of the planet Earth. In this connection, we also mention the books of McIlveen (1992), Peixoto and Oort (1992) and Panchev (2003).

The study of the processes and the phenomena in the ALO-system goes in two main directions – instrumental observations (in situ and remote sensing) and theoretical (mathematical modeling on physical basis). Now, in the era of space satellites and powerful computers, both of them undergo unprecedented progress. Hereafter, we discuss some particular aspects in the development of theoretical studies.

1.2 The Weather

Under the term weather one usually understands the state of the low atmosphere (the "kitchen" of the weather) at the moment or for a short period of time, over a limited area. This state is characterized by a set of meteorological fields (or elements) which are functions of the space-time coordinates (x, y, z, t), such as: wind velocity, temperature, pressure, humidity, precipitation etc. Very often only one of them is used as main characteristic of the weather – people say: it is warm, cold, windy, rainy, etc.

One of the most important scientific and practical problems is the understanding of the laws governing the weather and their application for its forecasting. No doubt that these are fundamental laws of the classical physics of continuous media – hydrodynamics, thermodynamics, radiation, water phase transitions, etc., with extensive use of mathematical tools.

A century ago, in 1905, V. Berkness formulated the weather forecasting problem as a mathematical one for solving the system of partial differential equations (PDE) of the atmospheric thermo-hydro-dynamics at given initial and boundary conditions. Two decades later, L.F. Richardson realized (by hand!) the first (unsuccessful!) such a numerical weather forecast. Still 20 years later, the first electronic computer (ENIAC) was used by J. von Neumann for this purpose.

Meanwhile and next, the physical basis of the mathematical prognostic models was improved; the numerical methods for solving the equations were further developed; the volume and accuracy of the observational data gathered in the world set of meteorological stations and by remote sensing (i.e., by space satellites) methods increased many times and finally, the computers became much more powerful in memory and speed. It was logically to expect that the quality and duration of the meteorological forecasts will also increase proportionally. However, this was not the case!

An improvement was actually observed, but with a clear tendency towards "saturation" – the results did not correspond to the material and intellectual resources engaged into the problem. In other words, a "horizon of predictability" for deterministic forecasts exists. It was theoretically evaluated as 2 weeks and for now it has practically reached 1 week. There are at least five reasons for this:

– imperfect physics under the mathematical prognostic models;
– observational and other errors in the initial data;

– unknown boundary conditions for the regional models over limited area;
– inappropriate numerical schemes for discretization of the equations;
– the equations themselves – nonlinear partial differential, large number, multiparametric, etc.

The fundamental explanation, however, came after the 1960s from the theory of chaos – the crown of the classical physics (Panchev 2001). The ALO-system is nonlinear and extremely complex with a great number of feedbacks. Even the simplest mathematical models must be nonlinear. But such completely deterministic equations with parameters can generate highly irregular, nonperiodic, stochastic-like solutions with exponential sensitivity to the initial conditions, so that errors in the latter will grow very rapidly during the model runs. In other words, we have what is called "chaotic solutions". After some "term of predictability", the errors become large enough to completely compromise the forecast. The understanding of this fact results in the quite pessimistic conclusion – an ideal and unlimited in duration deterministic numerical prognosis could never be achieved even if most of the above mentioned factors were eliminated. This would be possible to some extent if there were some highly predictable effective reasons governing the weather. But this is not the case! That is why, the idea to use the solar activity is so debatable. Most probably, the ALO-system has its internal causes for the irregular weather variability and limited predictability.

During the last decade, our research interest is in the field of general theory of nonlinear (chaotic) dynamical systems and their various applications, including meteorological ones. A number of Bulgarian scientists and groups, mainly in the frames of international projects and programmes, have contributed to a better understanding of the planetary boundary layer physics and dynamics, the ocean dynamics, etc. This allows more correct formulation of the low boundary conditions in the numerical weather prediction described above. Now, several weather forecast models are in operational use in the Bulgarian hydrometeorological service – ALADIN, ARPEGE, MM5, etc. Results from other models are also used, such as the model of the European Centre for Medium-Range Weather Forecasts. Some of these models (ALADIN, for instance) have regional versions with smaller space and time steps, which better accounts for local terrain peculiarities and conditions, and allows to improve the weather forecast for the country.

1.3 The Climate

There are many definitions of climate in the popular literature. The physically rigorous definition for climate reads: "The climate of the Earth is a statistical ensemble from the states of the system ALO (called also Climatic System (CS)) for a long enough period of time τc". It is recommended by the World Meteorological Organisation (WMO) $\tau c = 30$ years because for such a period, the mean values of the climatic fields are relatively stable and representative. For time intervals $\tau < \tau c$, one speaks for climatic variability, while at $\tau > \tau c$ – about climatic changes (tendencies, trends). Moreover, the above definition permits to speak about "regional" and "local" climate, "climate of the ocean", "space weather and climate", etc. (Panchev 2003). In these cases, the interactions with other parts of the system must be considered. Evidently, it is not possible to distinguish sharply short-period ($\sim$1 year) climate variation from long-period weather fluctuations and for this reason, terminological mixing exists in the practice.

What is the role of physics in the problem of climate change? Quantitative diagnosis of the climatic changes during the geological history of the planet is impossible without knowledge of physical laws. Instrumental measurements have been made during the last few centuries only due to the physical and engineering achievements. Physical (radioisotope) methods ensure reliable data for climate description before this period. On such basis only, an acceptable explanation of the real climatic changes can be searched for. It is well known that long and short, irregular in time, glacial and interglacial periods have happened. Their "signatures" have been seen on the land, ocean bottom and glaciers, including the Antarctic ones.

The knowledge and the understanding of the climate history is a precondition for development of a reliable theory capable to predict climate future. In essence, the theory of climate is a statistical dynamics of the CS and is one of the most challenging theories at present. Two basic goals in development of this theory can be pointed out:

- Reproduction of past events in the climate history with or without inclusion of cosmic factors, i.e. solar activity;
- "Prognosis" for the future climate of the Earth accounting for new factors such as the anthropogenic greenhouse gas emissions. Here, the inverted commas stand to stress on the new meaning of the word prognosis concerning the climate consistent with the definition. One has to talk and write about climate perspectives with some probability or statistical moments, instead of climate prognosis.

Respectively, the modern climate theory goes in two main directions:

- Analysis and interpretation of the real climatic data;
- Development of mathematical models on physical background and numerical experiments with these models.

The first direction aims at identification of the physical and other factors influencing the climate and their quantitative representation. The second one aims at bringing the factors together in equations which is not unique operation. This is the reason for existence of a wide spectrum of models containing from one to several dozens of equations depending on the purpose of the model and the resources for its realization. The simplest, so called "point models", consist of up to three nonlinear ordinary differential equations of evolutionary type $da/dt = \dot{a} = b - c$. Two examples for such models are given in the Appendix (see also Panchev 2001). They tie globally averaged quantities such as temperature, CO_2 content, ice mass, etc. and are comparatively simple for solving and analysis. The "heavy" climate models, such as global circulation models, describe the CS as a whole and consist of large number PDEs. Only several national or world meteorological centres operate with such models (American, European, Canadian, Russian, etc.) because they require powerful computers, abundant collection of data as well as considerable intellectual and financial resources.

The problem for "prediction", or more precisely "perspectives" of the climate for decades or 1–2 centuries ahead is already a problem not of physics (geophysics) only but multidisciplinary one. It has demographic and economic aspects related to the growth of human population on the planet and the increasing influence of the anthropogenic factor. The physical aspect is, to the great extent, understood and mastered. As to the other aspects, they are presented in the models in the form of scenarios, i.e. assumptions on how respective components will change in the future – staying uncontrolled or being limited by international protocols and agreements. Under such situation, various climate change models can be created accounting for both type of aspects and predicting the reaction of the CS depending on the initial data. Preliminary results concerning the future climate up to the end of the twenty-second century have been recently published (Dai et al. 2001a, 2001b, 2001c). The authors conclude that global warming of about 2–3°C could be expected, against less than 1°C increase at the end of the twentieth century, compared to the nineteenth one.

1.4 Conclusions

In the light of the above, a fundamental question arises: when and to what extent one can believe to the model perspectives for the climate? From theoretical considerations, the answer is: when the model can satisfactory reproduce past events from the climate history (McIlveen 1992). It is found that only internal mechanisms of the CS can explain qualitatively its observed behaviour. Moreover, the models predict more than one stationary climate, different from the present one, so that the CS can "jump" from one to another stationary state if the physical parameters of the system change above some threshold value (Monin and Shishkov 1979).

An important feature of the climate modeling research is the possibility for numerical (computer) experimentation. In such experiments, one can start by giving various initial conditions like absence or different configuration of the continents and oceans on the globe, different solar radiation by some reasons, different composition of the atmosphere, absence of biosphere, etc. Different climates will correspond to each of these cases. Hence, the life on the planet will react correspondingly.

Finally, the problem for detecting regional and local manifestations of the global climatic changes is not less difficult. Technically, it is similar to the problem to distinguish and separate the useful "signal" on the background of the "noise" of same order. This important problem, however, can be attacked by single scientists or small research groups. Bulgarian physicists-meteorologists also contribute to find solutions of this problem with regard to the climate in the region of South-Eastern Europe, the Balkan Peninsula and Bulgaria in particular.

As to the weather forecast models, all of them are being constantly improved and behave very well even for limited area and for a period from 1 week up to 10 days. However, this is correct for relatively stationary synoptic situations. In case of more dynamic non-stationary situations, the accuracy of these models is usually worse. For this reason, the extreme events are much more difficult to be exactly predicted in space, time and intensity. The development and improvement of the weather forecast models in this sense have to be encouraged.

Appendix

Below, examples for model equations of two dimensionless point climatic models are given. Even very simple, they could demonstrate quite rich spectrum of solutions' behaviour, respectively – quite different climates could result as their solutions.

(1) The Lorenz model (Lorenz 1984, 1990)

$$\begin{aligned} dX|dt &= Y^2 - Z^2 - aX + aF, \\ dY|dt &= XY - bXZ - Y + G, \\ dZ|dt &= bXY + XZ - Z, \end{aligned} \tag{1.1}$$

where $X(t)$, $Y(t)$ and $Z(t)$ represent the zonally averaged meridional temperature gradient (or equivalently, the intensity of the westerly wind current) and the amplitudes of the cosine and sine phases of a chain of superposed large scale eddies, respectively; aF and G stand for the symmetric and asymmetric external thermal forcing by the underlying earth surface (land and oceans); a and b are physical parameters.

(2) The Saltzman-Maasch model (Saltzman and Maasch 1988)

$$\begin{aligned} dX|dt &= -X - Y, \\ dY|dt &= -pZ + rY - sZ^2 - YZ^2, \\ dZ|dt &= -q(X + Z), \end{aligned} \tag{1.2}$$

where $X(t)$, $Y(t)$ and $Z(t)$ represent the global ice mass, the CO2 content and mean static stability of the world ocean (i.e. the Brunt-Vaisala frequency) whereas p, q, r, s are model parameters.

References

Dai A, Meehl GA, Washington WM et al. (2001a) Ensemble simulation of 21st century climate changes: Business-as-usual versus CO_2 stabilization. Bull Amer Meteorol Soc 82(11): 2377–2388.

Dai A, Wigley TML, Boville BA et al. (2001b) Climates of the 20th and 21st centuries simulated by the NCAR climate system model. J Climate 14:458–519.

Dai, A, Wigley TML, Meehl GA et al. (2001c) Effects of stabilizing atmospheric CO2 on global climate in the next two centuries. Geophys Res Letters 28(23):4511–4514.

Lorenz EN (1984) Irregularity – a fundamental property of the atmosphere. Tellus 36A: 98–110.

Lorenz EN (1990) Can chaos and intransitivity lead to interannual variability. Tellus 42A:378–389.

McIlveen R (1992) Fundamentals of Weather and Climate. Chapman and Hall, London.

Monin AS (1969) Predskazania vremia kak problemma fiziki (The Weather Forecasting as Problem of Physics). Nauka, Moscow.

Monin AS, Shishkov UA (1979) Istoria klimata (The history of Climate). Gidrometeoizdat, Leningrad.

Monin AS, Shishkov UA (2000) Klimat kak problemma fiziki (The Climate as Problem of Physics). Uspehi Physicheskih Nauk 170:419–445.

Panchev S (2001) Teoria na haosa (Theory of Chaos), 2nd edn, Bul Acad Press, London.

Panchev S (2003) Osnovi na atmosfernata fizika (Foundations of Atmospheric Physics). Bul Acad Press, London.

Peixoto JP, Oort AH (1992) Physics of Climate. American Institute of Physics, New York.

Saltzman B, Maasch KA (1988) Carbon cycle instability as a cause of the late pleistocene ice age oscillations: modeling the asymmetric response. Global Biochemical Cycles 2(2):177–185.

Chapter 2
Climate Change and the Balkans: Real Concern or "Useless Arithmetic"

C. Gregory Knight

Abstract The theme of this essay is the importance of a focused Balkan regional analysis of future climate change and climate change impacts on a more detailed, cross-national and regional basis than has been undertaken before. The argument is based on the differences in potential future climate in the Balkans compared to the European continental area in which the region is typically included and on the reality that geographical propinquity may be more important than boundaries of nation states when issues of climate change impacts are considered. Differences among climate change scenarios and uncertainty in resulting impact patterns are not grounds for ignoring the importance of regional collaboration on climate change assessment. Thus the creation of an informal collaborative among Balkan climate change scientists is proposed.

Keywords Balkans · Climate change · Impacts · Regional assessment · Cooperation

2.1 Introduction

The Intergovernmental Panel on Climate Change process, with its history of four increasingly sophisticated climate change documents and additional special reports, has stressed the importance of examining the potential impacts of global climate change with a focus at the continental scale.[1] Regional investigations at the scale of the European Union have national and sub-national resolution, and fortunately include non-EU members in the Balkans.[2] In 2002, *GeoJournal* published a special

C.G. Knight (✉)
Department of Geography, Pennsylvania State University, University Park, PA, USA
e-mail: cgk@psu.edu

[1] See http://www.ipcc.ch for full documentation of the IPCC reports.

[2] See http://ec.europa.eu/environment/climat/home_en.htm as an entry to EU-level climate change documents.

V. Alexandrov et al. (eds.), *Global Environmental Change: Challenges to Science and Society in Southeastern Europe*, DOI 10.1007/978-90-481-8695-2_2,

issue on climate change in central and eastern Europe with reviews of some relevant activities in the region, including a summary of national activities up to that time (Knight et al., 2002; Knight and Staneva 2002). Projects reported in other chapters of this volume include selected Balkan nations and research sites, along with non-Balkan countries whose prognoses of potential climate futures and impacts vary widely from those of Balkan areas. Various other climate change programs, from the US Country Studies Program in the 1990s (Smith and Lazo 2001) to the on-going activities and reports of the UN Framework Convention on Climate Change,[3] focus on national level assessments with weak linkages, if any, beyond national boundaries (exceptions, of course, relate to impacts transmitted by hydrological impacts in shared international basins).

The work presented in this symposium points strongly to the opportunity and indeed necessity for attention at the level of the Balkan Peninsula. In this essay I would like to elaborate this contention, drawing upon the concept of integrated regional assessment (IRA, elaborated below). Such an assessment for a region like the Balkans could lead to synergies among researchers in nations tied together by natural processes as well as shared histories and futures. This chapter, then, is not a report on climate change science or an impact assessment. It is intended as a call for collaboration across the Balkan Region to enhance understanding of future opportunities and threats, as well as of potentials for climate change mitigation and adaptation. Resilience in the face of challenge may well depend on cooperation and collaboration across national borders that are artifacts of history and not necessarily significant environmental boundaries.

My argument and proposal begins with a brief discussion of the "integrated regional assessment" (IRA) concept, turning subsequently to the climate scenarios that are a core element of such activities. Here we ask the question: if climate cannot be predicted and if various attempts to do so have results at variance from one to another, are climate scenarios simply "useless arithmetic?" I will argue that is not the case, based on understanding of differences among predictions, forecasts, projections and scenarios, bolstered by recognition that the precautionary principle may well require action based on uncertainties that will hopefully become more well-defined as both science progresses and the future unfolds. Unfortunately, by the time the region and the world as a whole see with certainty what climate change is doing, trajectories toward disaster may be unavoidable (Giddens 2009). What, then, could be learned from an integrated regional assessment of climate change in the Balkans? This question is explored with reference to the meaning of integrated regional assessment and how this process could be integrated into policy and planning in the region.

---

[3] See http://unfccc.int for details and access to extensive documentation on the emissions, mitigation and adaptation reports of individual nations.

2.2 Integrated Regional Assessment

There have been many global analyses of climate change and impacts, the most prominent of which have been summarized by the Intergovernmental Panel on Climate Change in its four reports issued to date (IPCC 2007). Similarly, there are studies at the European scale, an example being the Eurowasser assessment of climate impacts on water resources (Lehner et al. 2001). At least two projects addressing central and eastern Europe are presented in this book, CLAVIER (Chapter 10) and CECILIA (Chapter 11). All these activities have varying degrees of spatial resolution, make use of various global socio-economic scenarios used by the IPCC and draw upon different ensembles of general circulation model (GCM) results. The GCM results are not an end in themselves, but a part of a larger process of assessment of climate change impacts, most frequently at a sub-global, regional scale.

There also exists an extensive literature on integrated regional assessment, including many exemplary studies and a summary volume on IRA (Knight and Jäger 2009). *Integrated* refers to the use of interdisciplinary knowledge; *regional* refers to the special resolutions from local to state, nation and region in which people largely interact; and *assessment* refers to the application of science to significant policy issues. IRAs may originate from the scientific community bringing important issues to the attention of policy makers; they may arise from concerns of policy makers for guidance on issues they are addressing. Using terms from Pielke (2007), some of the underlying science may come from the work of *pure scientists* working oblivious of policy issues and others from the work of *science arbiters* who address their work to answering challenges from policy makers and stakeholders. A fair amount of the science will come from *issue advocates* (and will be contradicted by others who take a skeptical viewpoint). Preferably, policy will be guided by *honest brokers* of science, scientists who translate and clarify their relevant work for the policy community.

IRAs are both a process and product. They are a process of interaction among scientists, policy makers, stake holders and others who interact to guide the evolution of a specific project; they are a product in the form of documents and websites from which findings are disseminated beyond the assessment team to wider publics.

It is common that an IRA begins with documentation of the current status and trends in a region, followed by analysis of the threats and opportunities under climate change scenarios. It will address the vulnerability of the region to climate, including potentials for adaptation and specification of the levels of resilience in natural and human systems. Since the IRA is strongly driven by assumptions about global greenhouse gas emissions, which in turn are major forcing functions in climate models, there are initial socio-economic uncertainties even before any model is run or impacts suggested. Impacts are identified themselves through models driven by GCM outputs at relevant scales in time and space, by expert judgment, by reference to analogous situations in the past or by reference to places with present climate that is analogous to climate scenarios for the region in question. Important outcomes of IRAs are a sense of the degree of anticipated changes and the time frame in which environment and society will be called upon to respond.

Let us turn, then, to the important role of climate scenarios in the IRA process, recognizing of course that the global trajectories of emissions, energy use, economy, population and other factors precede the application of GCMs for assessment purposes.

2.3 Regional Climate Scenarios

Important dimensions of all the climate change assessment activities mentioned are the very different conditions and trajectories anticipated for the Balkan/Mediterranean area vs. continental northern and eastern Europe. For example, the latest IPCC regional analysis of climate models shows a high degree on consonance in anticipated changes in this region, compared to ambiguity in parts of Europe within which the Balkans have often been included. If one looks at the preliminary scenarios of CECILIA, regional trajectories are for increasing temperature and increasing precipitation, a conclusion strongly at variance for scenarios for the Balkans. Some examples of climate projections and impacts for the Balkan Region are presented in other chapters in this volume. IPCC summarizes European climate change scenarios as follows:

> Annual mean temperatures in Europe are *likely* to increase more than the global mean. Seasonally, the largest warming is *likely* to be in northern Europe in winter and in the Mediterranean area in summer. Minimum winter temperatures are *likely* to increase more than the average in northern Europe. Maximum summer temperatures are *likely* to increase more than the average in southern and central Europe. Annual precipitation is *very likely* to increase in most of northern Europe and decrease in most of the Mediterranean area. In central Europe, precipitation is *likely* to increase in winter but decrease in summer. Extremes of daily precipitation are *very likely* to increase in northern Europe. The annual number of precipitation days is *very likely* to decrease in the Mediterranean area. Risk of summer drought is *likely* to increase in central Europe and in the Mediterranean area. The duration of the snow season is *very likely* to shorten, and snow depth is *likely* to decrease in most of Europe.[4]

What can we say at the highest level of generality about climate scenarios for the Balkan Region, drawing upon the last IPCC report? A useful perspective is to ask most fundamentally about the directions of climate change that could emerge, and then of course the magnitude, timing, and variability of change resulting from GCMs. For its latest report, the IPCC considered nineteen GCMs, results of which were summarized in various maps suggesting directions of climate change and the proportion of GCMs which concurred on those directions of change (IPCC 2007).

Here five dimensions of modeled change are addressed, in respect to the Balkan Region – summer and winter temperatures, summer and winter precipitation, and

---

[4] http://www.ipcc.ch/pdf/assessment-report/ar4/wg1/ar4-wg1-chapter11.pdf; last accessed 10 October 2009

annual runoff. This is not to ignore other issues such as variability, persistence of wet and dry periods, growing season length and the like, but rather to paint a broad-brush picture of changes that could occur across the Balkans and to suggest that for that region, "we all are in it together" and the issues loom large, not just for nations but the region as a whole. Of course this is not to say that with increasingly sophisticated methods of downscaling from the rather coarse geographic grids of GCMS to regional and local levels, nuances of local climate control and other local changes such as land use will give texture to a wider regional picture. In earlier research, we suggested that some GCM results for Bulgaria projected "normal" climates by the end of the 20th century that could be drier than the severe drought that occurred in Bulgaria and adjacent countries in the 1980s through the mid 1990s (Knight et al. 2004). Even with the relatively simplistic downscaling used at that time, it was apparent that there would be regional differences even within one country, albeit in degree of drying, not direction of the change toward aridity.

Using IPCC maps of the five variables mentioned, it seems there is great consistency (more than 90% of the 19 models) in saying that both winter and summer temperatures could increase significantly, that summer precipitation could decrease by 10 to over 20%, also with results consonant over 90% of GCMs. No consistent signal emerged from the GCMs for winter precipitation. The directions of these patterns appear to be continuous across the Balkan Peninsula, extending south into Greece, east into Turkey and north into Romania and Moldova.

Runoff represents the complex interaction between climatic dimensions of precipitation and the atmosphere-surface energy regime. Runoff can decrease due to less precipitation, more evaporation, and/or changes in the timing of events and seasons. The IPCC map of modeled changes in runoff clearly shows sharp and significant decreases in annual runoff across the Balkans on the order of ***minus*** 20–40%, yet again with greater than 90% agreement among GCMs. There will be local differences in sensitivity to such decreases, of course, but logical consequences for such changes include threats in agriculture, irrigation, hydroelectric power, and even potable water during what could be "normal" times, and even more so during inevitable dry excursions from future average or normal climate.

In looking at these first order generalizations, it is reasonable to argue that perhaps world trajectories of greenhouse gases may not take IPCC-specified paths and change might not be so threatening. One would have to be an accomplished optimist, however, to assume that global climate will not change at least to some degree in ways that bring the kinds of trajectories cited, and that superimposing climate variability on top of these changes could mean that today's threatening events become catastrophic in future. Indeed, there seems to be indication that climate change has been accelerating since the last IPCC report, and that in future IPCC scenarios representative concentration pathways (RCP) of high greenhouse gas accumulation once considered unlikely may be moderate in comparison to others (Moss et al. 2008). Nevertheless, if all we can generate are scenarios, and not predictions, and the scenarios are rife with assumptions and uncertainties, why heed them?

2.4 "Useless Arithmetic"?

The phrase "useless arithmetic" is borrowed from the title of the 2007 book by coastal scientist Orrin Pilkey and his daughter, Linda Pilkey-Jarvis. The book's subtitle is "Why environmental scientists can't predict the future." In their book, the authors argue that the complexity of realty is so great that no systems model can predict the future. Thus, they argue, the best scientists can do is give qualitative judgments about possible futures, given what they refer to as "... virtual impossibility of accurate quantitative modeling to predict the outcome of natural processes on the earth's surface."

Factors cited by Pilkey and Pilkey-Jarvis include errors in characterizing processes, omission of important processes, inadequate knowledge of initial conditions, forces not included in the modeling system, scale effects, and non-linearities. We know that scientists and engineers use models all the time in the design process – for airplanes, dams, flood controls, traffic flows – and that the systems they design are, overwhelmingly, successful. That said, it is still reasonable to be skeptical about models that extend for decades in the future?

Given the indeterminacy of socio-economic process that are driving climate change, one could agree with the Pilkeys that climate change models may be "useless arithmetic," particularly if policy-makers who would use the work of climate scientists are overwhelmed by uncertainties: the noise would seem to drown out the signal! Box and Draper once wrote:

> Remember that all models are wrong; the practical question is how wrong do they have to be to not be useful (Box and Draper 1987: 74)

So how do we address the dilemma of model uncertainty? Part of the issue is rooted in the confusion among predictions, forecasts, projections, and scenarios.

2.5 Predictions, Forecasts, Projection and Scenarios

In our book on regional assessment of climate change, Knight and Jäger write, following MacCracken (2001), that

> ... we need to emphasize that integrated regional assessments are *not* predictions of the future. It is important to make a clear distinction between the terms prediction, forecast, projection and scenario... A prediction is a statement about the future based on current conditions and reasonably well-specified methods of anticipating the future. A forecast connotes prediction whose credibility depends on the person doing the predicting; it also implies a time frame in which the forecasts can be tested and refined. In contrast to both prediction and forecast, a projection is based on one of several or many initial conditions and specification of how those conditions may change in the future. In most climate change assessments, there are a number of initial assumptions and future trajectories of driving forces and mitigation actions, all used as input to a variety of different climate models. Such projections create climate scenarios, neither predictions nor forecasts, as bases for assessment of what *could* happen in future. General circulation models of the atmosphere are retrodictions when they account for the climate record; for the future they generate projections, which can be summarized as scenarios; they are decidedly not predictions. Derivation

> of impacts from these scenarios is similarly of scenario, not prediction, status, whether the derivations come from loosely linked qualitative analyses or structured mathematical models. Thus the output of an integrated regional assessment process is neither a prediction nor a forecast. Rather it is a plausible set of scenarios that could occur under specified conditions (Knight and Jäger 2009: 10–11).

Climate investigations for the Balkans are not "useless arithmetic." By providing bounds on possible future climate within alternative global futures, policy-makers will have valuable insights for future planning. National policies that will extend into the future, including those related to structural investments of five decades or more of technical life, will be better informed in view of the range of climate futures than relying on past environmental measurements. We must question the assumption of the stationarity (constancy) assumption underlying probability distributions of weather and hydrological events for water resource engineering (Milly et al. 2008). Among additional future issues that should be considered are whether some existing climates will disappear and new ones emerge (Williams et al., 2007), calling into question whether a region's future climate has a contemporary analog, and how internationally shared environmental resources (especially water) might be managed.

2.6 A Strategy for a Balkan Integrated Assessment

It can be argued that collaboration across the Balkans could provide important perspectives for policy-makers as well as the scientific community. A suite of climate scenarios specified for the region could be helpful to all. These scenarios should provide methods for localized down-scaling and provide for frequent updating as the quality of GCMs approved. International collaboration on shared and boundary water basins, mountains, and human activities such as transportation, trade, and urban dynamics are among many opportunities to capitalize on scarce but high quality scientific resources across Balkan nations.

2.7 Conclusions and Suggestions

In conclusion, from the volume of activities already underway at the national and regional level, from expertise shared across the Balkans (as demonstrated by papers in the symposium, some of which are reported in this volume), from the interest of extra-Balkan nations and institutions in Balkan futures, and with clear evidence of shared patterns of potential climate change and impacts in the face of other common challenges, the potential benefits from a pan-Balkan integrated assessment of climate futures and impacts is obvious.

Several initiatives could lead to such an activity. First, a self-organizing group of international scholars could create an open organization linked by a web site and blog to share research and policy initiatives among Balkan communities. The core of such a network would include Slovenia, Croatia, Bosnia-Herzegovnia, Montenegro,

Albania, FYROM, Kosovo, Serbia, Romania, Moldova, Bulgaria, Greece and Turkey, roughly co-terminus with The South East European Co-operation Process (SEECP; Republic of Bulgaria. Ministry of Foreign Affairs 2009), not to the exclusion of interested colleagues worldwide. Hopefully, one of the regional institutions represented in this book will take a lead in establishing such a network. A Balkan regional assessment activity could emerge from such an activity. Second, a periodic, ad hoc, low-expense gathering could be hosted among participants in different locations on an annual or biennial basis to enhance collaboration via face-to-face discussion. Third, interested individuals from within the evolving Balkan climate change community and more distant collaborators should remain alert for funding opportunities that could contribute to achieving the longer term goal of an on-going assessment process for the region.

Acknowledgments I would like to acknowledge the contributions of members of the Scientific Coordination Center for Global Change (Bulgarian Academy of Sciences), the former Center for Integrated Regional Assessment at the Pennsylvania State University, and various individuals from around the Balkan Region and eastern Europe to my thoughts expressed here as they developed over time. My research in the region has been supported by the U. S. National Science Foundation, the International Research and Exchanges Board, and the Pennsylvania State University. Opinions, of course, are mine alone.

References

Box GEP, Draper R (1987) Empirical Model-Building and Response Surfaces, Wiley, New York.

Giddens A (2009) Politics of Climate Change, Cambridge Polity Press, Cambridge.

IPCC (2007) Climate Change 2007: Synthesis Report. Contribution of Working Groups I, II and III to the Fourth Assessment Report of the Intergovernmental Panel on Climate Change [Core Writing Team Pachauri RK and Reisinger, A. (eds.)]. IPCC, Geneva, Switzerland.

Knight CG, Jäger J (2009) Integrated Regional Assessment of Climate Change, Cambridge University Press, Cambridge.

Knight CG, Raev I, Staneva MP (2004) Drought in Bulgaria: A Contemporary Analog of Climate Change. Ashgate Studies in Environmental Policy and Practice. Aldershot, UK.

Knight CG, Staneva MP (2002) Climate change research in central and eastern Europe. GeoJournal 57:117–137.

Knight CG, Staneva MP, Hristov T, Balteanu D (2002) Climate change in central and eastern Europe: Introduction to a special issue of GeoJournal. GeoJournal 57:113–115.

Lehner B et al. (2001) EuroWasser: Model-based assessment of European water resources and hydrology in the face of global change, Kassel: Center for Environmental Systems Research (http://www.usf.uni-kassel.de/usf/archiv/dokumente/kwws/kwws.5.en.htm; consulted 16 September 2009)

MacCracken M (2001) Prediction versus projection – Forecast versus possibility, WeatherZine 26 (http://sciencepolicy.colorado.edu/zine/archives/1-29/26/guest.html; consulted 16 September 2009)

Milly PCD et al. (2008) Stationarity is dead: Whither water management? Science 319:573–574.

Moss, R et al. (2008) Towards New Scenarios for Analysis of Emissions, Climate Change, Impacts, and Response Strategies. Intergovernmental Panel on Climate Change, Geneva.

Pielke RA (2007) The Honest Broker. Making Sense of Science in Policy and Politics, Cambridge University Press, Cambridge.

Pilkey OH, Pilkey-Jarvis L (2007) Useless Arithmetic, Why Environmental Scientists Can't Predict the Future, Columbia University Press, New York.

Republic of Bulgaria, Ministry of Foreign Affairs (2009) SEECP (The South East European Co-operation Process). (http://www.mfa.bg/en/index.php?option=com_content&task=view&id=8089&Itemid=388; consulted 16 September 2009)

Smith JB, Lazo JK (2001) A summary of climate change impacts from the U.S. Country Studies Program. Climatic Change 50:1–29.

Williams JW, Jackson ST, Kutzbach JE (2007) Projected distributions of novel and disappearing climates by 2100 AD. PNAS 104(14):5738–5742.

Chapter 3
Global Environmental Change and Related Impacts

Lučka Kajfež Bogataj

Abstract Our common future is at risk because of global change, especially climate change. In this paper, we describe climate change scenarios based on high resolution climate models for Europe and the possible impacts of climate change due to extreme weather, natural disasters, desertification, and other climate events for different sectors including agriculture, water resources, biodiversity, human health, environment and infrastructure.

Keywords Climate change · Europe · Impacts · Agriculture · Natural disasters · Water resources

3.1 Introduction

Global change is a reality and a subject of enormous complexity linking large numbers of physical, chemical and biological factors in a planet-wide system. A wealth of scientific information, including paleostudies of past variability in the Earth's environment, observations of current changes, and model-based projections of the future evolution of the Earth's system, makes it clear that current changes are more rapid and profound than in past millennia and will continue for at least the next century. Humans have become a significant environmental force, most notably since the industrial revolution.

Increased population growth together with accelerated human activities and a rise in economic wealth over the past century have greatly increased resource use, which has been reflected in agriculture, fisheries, forestry, industry, transport, energy and urbanisation. This has resulted in multiple and interacting global environmental impacts, reflected in the current values of greenhouse gas emissions and rising temperatures. The primary cause of anthropogenic climate change is the gases and

L.K. Bogataj (✉)
University of Ljubljana, Ljubljana, Slovenia
e-mail: lucka.kajfez.bogataj@bf.uni-lj.si

V. Alexandrov et al. (eds.), *Global Environmental Change: Challenges to Science and Society in Southeastern Europe*, DOI 10.1007/978-90-481-8695-2_3,

aerosols that humankind's activities release into the atmosphere. These gases and aerosols influence directly the energy balance of our planet by trapping heat, and indirectly by changing, for example, the way clouds reflect sunlight. Over the next century, society will increasingly be confronted by the impacts of global change (e.g. pollution, land use changes, and climate change). Warming will shift climatic zones by intensifying the water cycle, affecting freshwater availability and human health.

Global change is evolving in very different ways in different regions of the world. The effects of global warming, for example, will be felt much differently at the regional level. Nearly all European regions are expected to be impacted in the future by climate change.

Central and Eastern Europe could face less summer rainfall, causing higher water stress. Health risks due to heat waves are expected to increase. Forest productivity is expected to fall and the frequency of peatland fires to increase. Southern European countries are very likely to face reduced water supplies, lower crop production, more wildfires and negative health impacts caused by increased heat waves. Northern countries are likely to benefit from increased crop yields, forest productivity, and supplies of food/fish from the North Atlantic. By 2020, most areas in Europe are likely to face an increased flood risk.

There is much evidence that – considering current climate change mitigation policies and related sustainable development practices – global greenhouse gases will continue to grow over the next few decades. A continuation in greenhouse gas emissions increase would cause further warming and induce many changes in the global climate system of the future that will be larger than those already observed. The impacts of climate change will vary greatly based on the development pathway of an area; estimates of regional population, income and technological development are strong determinants of vulnerability to climate change. Some future impacts already appear to be unavoidable, owing to the inertia of the climate system (Table 3.1).

3.2 Climate Change Scenarios Based on High-Resolution Climate Models in Europe

The prerequisite for intelligent, effective and efficient adaptation to climate change is a good understanding of regional impacts. Regional modelling of climate development is an essential basis to reach this goal. The use of different high-resolution climate models linked to impact models enables us to quantify the uncertainties of predictions and analyse how these uncertainties can be transferred from climate models into decision models. Major scientific progress has recently been made in this field.

PRUDENCE – a recent EU project (PRUDENCE 2005) – used four Atmosphere General Circulation Models (AGCM) and eight Regional Climate Models to quantify uncertainties associated with climate predictions and future impacts of climate change on Europe. PRUDENCE is providing improved model representation of

Table 3.1 Possible impacts of climate change due to changes in extreme weather and climate events, based on projections up to the mid to late twenty-first century (Modified after IPCC 2007b)

		Major projected impacts by sector			
Phenomena and direction of trend	Likelihood of future trend	Agriculture, forestry	Water resources	Human health	Industry/ settlement/ society
Fewer cold days and nights; more frequent hot days and nights over most land areas	Virtually certain	Increased yields in colder environments and decreased yields in warmer environments	Effects on water resources relying on snow melt; increased evapotrans-piration rates	Reduced human mortality from decreased cold exposure	Reduced energy demand for heating; increased demand for cooling; declining air quality in cities; effects on winter tourism
Warm spells/heat waves: frequency increases over most land areas	Very likely	Reduced yields in warmer regions due to heat stress; fire danger increase	Increased water demand; water quality problems, e.g., algal blooms	Increased risk of heat-related mortality, especially for the elderly, chronically sick, very young and socially-isolated	Reduction in quality of life for people in warm areas without air conditioning; impacts on elderly, very young and poor; reduced thermoelectric power production efficiency; disruption of commerce
Heavy precipitation events: frequency increases over most areas	Very likely	Damage to crops; soil erosion, inability to cultivate land due to water logging of soils	Adverse effects on quality of surface and groundwater; contamina-tion of water supply; water scarcity may be relieved	Increased risk of deaths, injuries, infectious, respiratory and skin diseases, post traumatic stress disorders	Disruption of settlements, commerce, transport and societies due to flooding; pressures on urban and rural infrastructure

Table 3.1 (continued)

Phenomena and direction of trend	Likelihood of future trend	Major projected impacts by sector			
		Agriculture, forestry	Water resources	Human health	Industry/ settlement/ society
Area affected by drought: increases	Likely	Land degradation, lower yields/crop damage and failure; increased livestock deaths; increased risk of wildfire	More widespread water stress	Increased risk of food and water shortage; increased risk of malnutrition; increased risk of water- and food-borne diseases	Water shortages for settlements, industry and societies; reduced hydropower generation potentials; potential for population migration
Sea level rise	Likely	Salinization of irrigation and well water	Decreased freshwater availability due to saltwater intrusion	Increased risk of deaths and injuries by drowning in floods; migration-related health effects	Costs of coastal protection, land-use relocation; displacement of human populations, abandonment of settlements, relocation of infrastructure

climate change scenarios by utilising high-resolution models (at spatial scales of ~50 km) for current (1961–1990) and future (2071–2100) climate scenarios, characterising the level of confidence in these scenarios, and assessing uncertainty resulting from model formulation. Future scenarios correspond to the IPCC A2 and B2 CO_2 emissions (IPCC 2001). Global and regional climate models (GC1Ms and RC1Ms) behave similarly in regards to temperature, except that GC1Ms exhibit a larger spread. The differences between GC1M and RC1M precipitation responses for some regions are significant. The spread of precipitation over the summer period is larger for RC1Ms than for GC1Ms. For both, however, the bias is twice as large as the response to climate change, when observed climate is used as a cross validation.

Scenarios indicate that European regions will undergo substantial warming in all seasons in a range of 1–4°C (B2 scenario) and 2.5–5.5°C (B2 scenario) by 2071–2100. Over Northern and Eastern Europe, warming has been stronger in winter, and the reverse happens over Western and Southern Europe, with stronger increases in summer temperatures (IPCC 2007a). Within Europe, future warming is estimated to be greatest over western Russia and southern countries (Spain, Italy, Greece), and less pronounced along the Atlantic coastline.

Across all scenario simulations, the results are in agreement on a general increase in winter precipitation in Northern and Central Europe and on a general decrease in summer precipitation in Central and Southern Europe, with the difference being a bit less marked in central Scandinavia (Raisanen et al. 2004). Over all, temperatures have seen an annual increase in Northern Europe and an annual decrease in Southern Europe. Increased Atlantic cyclonic activity could lead to stronger precipitation (up to 15–30%) in winter over Western, Central and Northern Europe, and anticyclonic circulation, in turn, to reduced precipitation in winter over Mediterranean regions in the south (Giorgi et al. 2004). In summer, a blocking situation caused by enhanced anticyclonic circulation over the northeastern Atlantic could lead to decreases in precipitation (up to 30–45%) over Western and Central Europe and the Mediterranean. Precipitation changes for spring and autumn are expected to be less pronounced than for winter and summer.

Notable changes are also projected for temperature and precipitation extremes in Europe. According to the IPCC (2007a), yearly maximum temperature is expected to increase much more in Southern and Central than in Northern Europe. Following EEA (2004a) projections, cold winters, which occurred on average once every 10 years over the period 1961 to 1990, are likely to become rare in Europe and will almost entirely disappear by 2080. In contrast, by 2080 nearly every summer in many parts of Europe is projected to be hotter than the top 10% of hottest summers in the current climate (EEA 2004a). Under high-emission scenarios, every second summer in Europe will be as hot or even hotter than 2003 by the end of the twenty-first century (Goodess 2005). In Southern Europe, these changes are projected to occur even earlier. This will be accompanied by unstable weather: events of extreme daily precipitation will increase even in most areas where mean annual precipitation decreases (Raisanen et al. 2004). The risk of drought is likely to increase in Central and Southern Europe.

Uncertainty in future precipitation projections is larger than those for temperature. This applies particularly to regional precipitation patterns and seasonal distribution of precipitation. But it should be stated that scientific confidence in the ability of climate models to estimate future precipitation is steadily increasing (IPCC 2007a).

In many impact studies, two global scenarios are usually selected from the IPCC Special Report on Emissions Scenarios (SRES), belonging to the A2 and B2 scenario storyline. These choices partly cover the range of uncertainty associated with the driving forces of global emissions: demographic change, economic development, and technological change. In the A2 scenario, where the storyline focus is on national economy, global greenhouse gas emissions are assumed to increase more significantly, leading to approximately a tripling of average CO_2 concentrations by the end of this century compared to pre-industrial concentrations. The B2 storyline focuses on local stewardship and results in approximately a doubling of the atmospheric CO_2 concentration. Those concentration levels translate into the following global mean temperature increases in 2071–2100, relative to 1961–1990: under scenario A2, an increase of 3°C, and under scenario B2 a temperature increase of 2.2°C.

3.3 Climate Change Impacts in Europe

Europe's natural environment, its production systems (agriculture, fisheries, forestry, terrestrial ecosystems) and other key socioeconomic sectors (tourism, energy, human health care, etc.) are under pressure from environmental change and socioeconomic development. Climate change creates additional pressure and the impacts of climate change on the environment and society are being observed across the region. This is already having an economic impact in Europe, particularly regarding recent extreme weather events. The following chapter discusses key natural environmental and societal impacts for Europe.

3.3.1 Natural Disasters, Water Resources and Desertification

Even without manmade climate change, society faces a wide range of natural disasters that can have serious social and economic consequences for people and their livelihoods. Within Europe, more than elsewhere, these threats are becoming greater as population densities increase. In addition, natural disasters can have negative environmental consequences, causing destruction to delicate habitats and ecosystems. In the future, extreme events induced by climate change – such as intense precipitation leading to flooding and landslides – are expected to increase in frequency and magnitude. In addition, poor land management practices combined with the effects of increased summer continental drying may lead to drought and widespread desertification, turning previously fertile land into badlands. It is important to understand where and how natural disasters might arise and what can be done to reduce their impact on people and the environment.

Since 1998, Europe has suffered more than 100 major floods, causing extensive damage. As illustrated by floods in the summer of 2005 and previous years, dramatic inundations are on the rise in Europe, with devastating impacts on human life, the economy and the environment. The massive flooding of Germany and Poland in 2002, for example, was the worst since 1845. Scientists predict this is only the beginning of the trend as climate change accelerates. England was affected by flooding in June and July 2007, Britain's wettest May–July since record counting began in 1776. Greater and more intense precipitation has already been observed in many areas of Europe, and this trend is expected to strengthen in the future. MICE's study of the projected climate in 2070–2099 suggests a reduction in intense rainfall over Southern Europe and an increase over the continent's northern region.

Flooding is not only on the rise due to climate change, however. The relentless canalisation of streams and waterways over the last 100 years has combined with urban exploitation of natural flood plains and marshlands to raise the stakes. Ground absorption rates have declined; runoff from melted snow and precipitation finds fewer and fewer natural channels. The result is an increasing incidence of flash flooding at the local level and catastrophic regional flooding.

According IPCC (2007c) estimates, a 4% increase in the frequency of heavy precipitation in the mid and high latitudes of the Northern Hemisphere over the second half of the twentieth century seems to have already taken place. Ironically, each year since 1990, the average land area and population affected by droughts has doubled. Water scarcity is a problem affecting at least 14 EU Member States and around 100 million inhabitants in 26 river basin districts throughout Europe. Based on the results of trend analysis for the last 40 years, long dry periods in summer increased in most parts of Central Europe, the UK and southern Scandinavia, and long dry periods in winter increased in Southern Europe.

Water is critical, and impacts have a cascade effect. Changes in water demand strongly depend on economic growth and societal development, as well as evolving needs in other sectors. Economic sectors projected to be most affected are (EEA 2007b): agriculture (increased demand for irrigation), energy (reduced hydropower potential and cooling water availability), health (worsening water quality), recreation (water-linked tourism), as well as fisheries and navigation. There are potentially serious impacts on biodiversity. Other serious effects could include flooding in Central Europe, lack of hydropower, health and ecosystem concerns in northern countries, and water scarcity in southern countries.

Alcamo et al. (IPCC 2007b) predict that the percentage of area under high water stress in Europe is likely to increase from 19 today to 35% by the 2070s, and the additional number of people affected by the 2070s is expected to be between 16 and 44 million. The most vulnerable areas will be those already water-stressed, and developing regions lacking water management systems which could act as buffers to increasing variability in water quality and quantity. There are only a few studies on the impacts and economic costs of these changes.

Several countries in Eastern Europe show signs of accelerated desertification: Hungary, Romania, Moldova, Bulgaria, Ukraine, and Southern Russia (Fig. 3.1), which all fall into the category of European areas with less precipitation. They

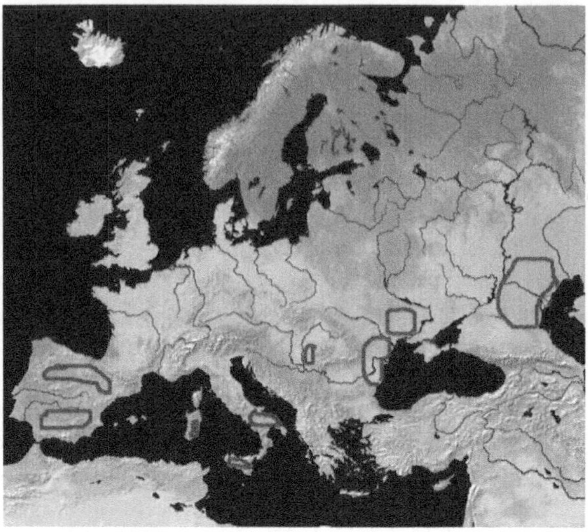

Fig. 3.1 Map of Europe with zones of desertification

have all experienced exhaustive land exploitation during the twentieth century. Economic mismanagement of agriculture in communist times was usually compensated by developing additional acres of land. Unlike Romania and Bulgaria, which are blessed with mountain ranges and thus able to store some water during springtime, Hungary, Moldova and Ukraine depend almost entirely on rivers. The need to divert rivers for industrial use not only affects agriculture, but also causes land to become less fertile. To sum up, traditionally low levels of precipitation plus water waste due to agriculture and industry all contribute to desertification. The consequences are several and uniformly negative for people living in these areas. Fertile lands will become less available, increasing economic pressure on remaining land, and accelerating the process of desertification. People will move permanently out of these areas, increasing migration pressure within and out of these countries. The urban population will have to import more expensive food from Europe, rendering locally produced goods more expensive and less competitive, which will further increase economic problems.

3.3.2 Agriculture

Agriculture is one of the biggest uses of European land, and as such is highly dependent on environmental conditions. Food production is still the major focus of agriculture and Europe is one of the world's largest and most productive suppliers of food and fibre (in 2004, 21% of global meat production and 20% of global cereal production). As European agriculture is highly intensive, weather remains the main source of uncertainty for crop yield assessment and crop management (Metzger et al. 2004).

The wide-ranging impacts of climate change have been documented in Europe over the last decades. For instance, the European heat wave of 2003 had major impacts on agricultural systems by decreasing the quantity and quality of harvests, particularly in Central and Southern Europe. The very high air temperature and solar radiation resulted in a notable increase in crops' water consumption. This, together with a summer dry spell, resulted in an acute depletion of soil moisture and lowered crop yields. Throughout Europe, the main sectors hit by extreme climate conditions were the green fodder supply, the arable sector, the livestock sector and forestry. Potato and wine production were also seriously affected. More than 26,000 fires were recorded in Portugal, Spain, Italy, France, Finland, Austria, Denmark and Ireland. It is estimated that some 70,000 hectares of forest (not including agricultural areas) burned. The summer of 2003 caused additional side effects, felt the next year in the form of soil erosion and flooding, winter sowing problems, and the budding of trees (COPA COGECA 2003).

Over the last few years the EU has financed several large research projects on regional climate modelling and impact assessment. In particular, some projects (PRUDENCE, PESETA) have produced high-resolution maps with projected changes in climate variables, such as mean temperature and precipitation, and projected impacts, e.g. agricultural yields. These maps are very useful for policy making and awareness raising purposes (Fig. 3.2). They illustrate what can be expected in Europe by the end of the century, according to the IPCC scenario, if no action is taken to reduce GHG emissions, and global mean temperature increases by about 3.4°C.

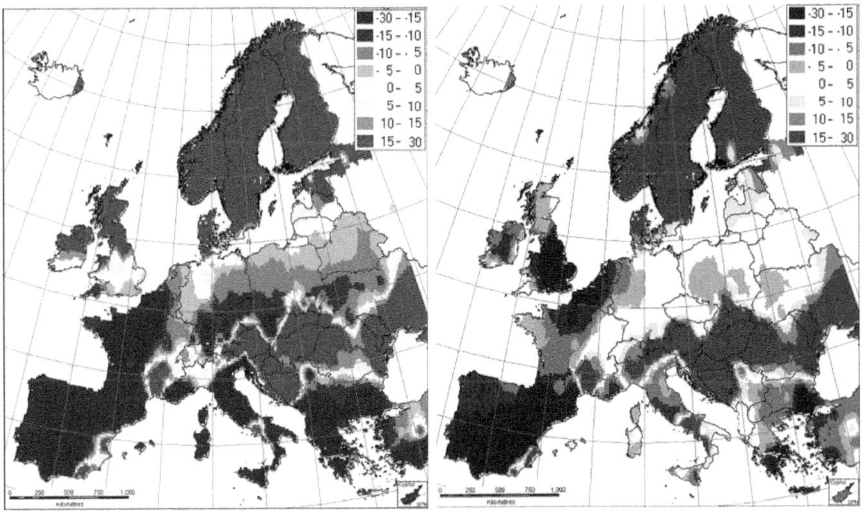

Fig. 3.2 Simulated crop yield changes (%) by 2080s according to two different models – (*left*) HadCM3, (*right*) ECHAM4 (European Commission 2007)

These maps are all based on IPCC SRES scenario A2. Results are impacts shown in the JRC-funded PESETA study (http://peseta.jrc.es). Maps with projections of future changes in temperature and precipitation are based on DMI/PRUDENCE data (http://prudence.dmi.dk), and processed by JRC within the PESETA study. Changes are projected for 2071–2100 relative to 1961–1990.

Climate change in Europe will very likely negatively impact ecosystem factors such as soil fertility, water availability, climate regulation potential and biodiversity, among others. Existing pressures on the already burdened natural environment of Europe may increase (Table 3.2). Furthermore, impacts may be unevenly distributed between the north (some positive, some negative) and south (nearly all negative). For example: the extent of forested land is expected to expand in the north and retreat in the south. Climate change will increase net primary productivity and total biomass in the north while reduced water availability is likely to decrease forest growth in Central Europe, and accelerate tree mortality in the south. On the other hand, the capacity of Europe's social systems to cope with climate change is high and is expected to continue rising. Adaptive capacity will vary between countries based on their different socioeconomic levels.

Changing climate, drier conditions and rising temperatures in the Mediterranean region and parts of Eastern Europe may lead to lower yields. Bindi and Moriondo (2005) predicted a general reduction in agricultural crop yields in the Mediterranean region, under IPCC SRES A2 and B2 scenarios by 2050, even when the fertilising effect of increased CO_2 is taken into account (Table 3.3). Similar yield reductions have also been estimated for Eastern Europe, with increased variability in yield, especially in the steppe regions (Maracchi et al., 2004).

Climate-related increases in crop yields are mainly expected in Northern Europe. For example, wheat yield increases are projected to be +2 to +9% by year 2020, +8 to +25% by year 2050 and +10 to +30% by year 2080 (Olesen et al., 2007; Audsley et al. 2006; Alexandrov et al., 2002). Another example is a sugar beet yield increase of 14–20% by the 2050s in England and Wales (Richter and Semenov 2005).

Uncertainties in the projection of future precipitation complicate estimates of future yield gains or losses. This is particularly true for Southern and South-Eastern Europe, where water will be a critical factor for agriculture in the future. In these areas, model results diverge to a great extent, depending on the scenarios in use and the model itself. For Central and Northern Europe, where water supply is less critical, projections are relatively robust.

Increased drought risk associated with global warming and a negative impact on water resources are among major concerns (Fig. 3.3). Several recent studies highlight the challenges that result from changes in water availability and water quality (EEA 2004b, 2005a,b; IPCC 2001, 2007c; Schröter et al., 2005). Under climate change conditions, it is expected that demands for irrigation will further rise, aggravating competition with other sectors whose demands for water are also projected to increase. In addition, an expected lowering of the groundwater table will make irrigation more expensive, which, in turn might have to be limited to cash crops. Extreme weather events such as heat waves will impact peak irrigation

Table 3.2 Some of the main expected impacts of climate change in Europe during the twenty-first century

Sectors and systems	Impact	Area: North	Atlantic	Central	Mediterranean	East
Water resources	Floods	—	—	——	–	——
	Water availability	++	++	+ to –	——	—
	Water stress	++	++	–	——	—
Forest, grasslands and shrub lands	Forest NPP	+++	++	++	+ to –	+
	Northward/ inland shift of tree species	+++	++	++	+ to –	—
	Natural disturbances (e.g., fire, insects)	–	–	–	——	—
	Change of stability of forest	—	–	–	—	——
Wetlands and aquatic ecosystems	Drying/ transformation of wetlands	—	–	–	—	——
	Disturbance of drained peat lands	——	NA	—	–	——
Agriculture and fisheries	Suitable cropping area	+++	++	+	—	–
	Agricultural land area	—	—	—	—	—
	Summer crops (maize, sunflower)	+++	++	+	——	—
	Winter crops (winter wheat)	+++	++	+ to –	—	+
	Irrigation needs	NA	+ to –	—	——	–
	Energy crops	+++	++	+	—	–
	Livestock	+ to	–	—	—	—
	Marine fisheries	++	+	NA	–	NA

Magnitude of impact: +, ++, +++. Type of impact: + = positive; – = negative; + to –: change in character through time, NA = not applicable.

Table 3.3 Changes of crop yields (%) for some Mediterranean regions by 2050 (modified from Bindi and Moriondo 2005)

	A2	B2	A2	B2
	Without CO_2		With CO_2	
C4 summer				
N-W	0.2	5.8	4.2	8,8
N-E	−4.4	−2.5	−0.6	0.2
Legumes				
N-W	−24.9	−13.4	−14.4	−4.9
N-E	−18.6	−8.1	−7.2	1.0
C3 summer				
N-W	−21.8	−10.4	−12.4	−2.9
N-E	−15.6	−6.9	−5.4	1.0
Tubers				
N-W	−10.4	−4.2	4.9	7.5
N-E	−22.5	−6.8	−9.3	4.4
Cereals				
N-W	−11.0	−3.5	−0.3	4.7
N-E	−6.8	3.7	4.4	12.5

N-W = Portugal, Spain, France and Italy; N-E = Serbia, Greece and Turkey

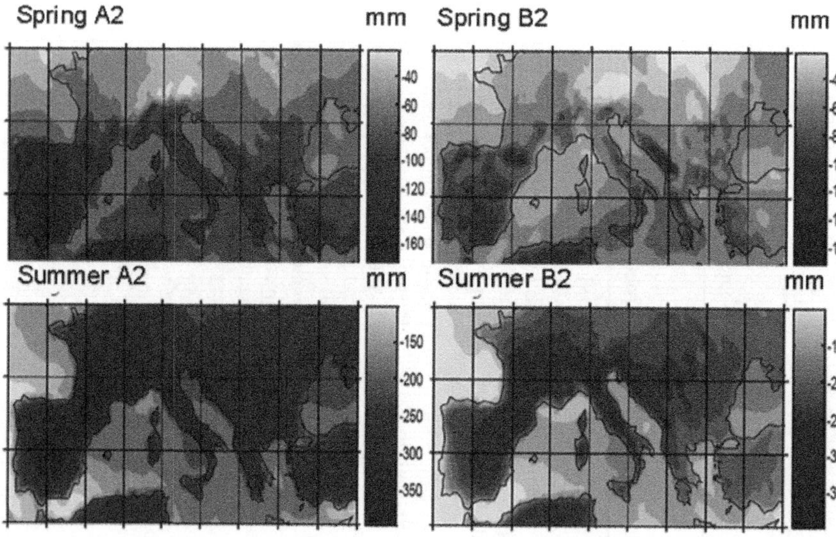

Fig. 3.3 Ensemble mean soil moisture changes in the Mediterranean between the periods 1961–1990 and 2070–2099 in spring and summer under the IPCC SRES A2 and B2 scenarios (PRUDENCE 2005)

requirements. As evaporative demand increases due to higher temperatures, it is expected that capillary rise will increase salinization of soils, having a major impact on irrigation management.

Overall, European food and fibre production is not expected to be greatly altered by climate change. However, greater differences will arise between countries. To sum up, it will be warmer in the north and drier in the south, rainfall will intensify, there will be an increased frequency in extremely hot days or seasons implying benefit to the north, and detriment to the south. This may worsen current resource issues: e.g., greater water shortages and heat stress in south, and more flooding in central, northern and mountainous regions. Furthermore it may aggravate current environmental problems (e.g., desertification in south; soil leaching in north). Taken together, the above implies a south-to-north geographical shift of climate resources in Europe, increasing the difference in resource endowment between the north and south of Europe. Policies might therefore be needed to support development in southern Europe, if the challenge of climate change is greatest there.

In the global context, Europe faces less negative effects than most other parts of the world, implying an opportunity to increase Europe's share of world food production. From an international viewpoint, it may be necessary to increase food production in Europe in order to maintain global food security.

3.3.3 Biodiversity

Biodiversity is the complex web of life on Earth and includes humans and our social and economic systems. Biodiversity also includes the genetic differences within each species – for example, between varieties of crops and breeds of livestock. Despite its overriding importance to our survival, biodiversity is being lost at a rate that is without precedent.

Biodiversity and ecosystems are important for maintenance and regulation of atmospheric quality (for example, they provide atmospheric oxygen and remove carbon dioxide), protection and cleansing of fresh water, marine productivity, soil formation and protection, erosion control, pollution breakdown etc. Human activity can upset the finely balanced interactions that keep ecosystems relatively stable. Human impact on ecosystems and climate change are already changing biodiversity worldwide.

Europe has a rich and varied biodiversity, which is feeling the pressure of expanding population, industrial technologies and transport, compounded by intensive exploitation of natural resources by industry, agriculture and fisheries.

Climate change will seriously affect biodiversity. Paleontological studies suggest that warmer phases in the Earth's history correlate with lower levels of biodiversity. If average global temperatures increase 3.4°C by 2100, it could cause the disappearance of 10–48% or more of the Earth's land surface.

Climate change presents specific dangers for the unique mountain ecosystem of the Alps. It is becoming very clear that some negative impacts of climate change can no longer be avoided. Warming in the Alps is anticipated to be twice as high as the European average.

3.3.4 Manmade Environment and Infrastructure

The manmade environment, including infrastructure, is also vulnerable to climate change, mainly through extreme events (floods and storms) as well as through increased heat waves and drought. The first two have a high potential for damage, while the latter are potentially important in relation to subsistence. Predicting the future effects of extreme events is difficult. First, there is less confidence in climate model predictions for such events. Second, exposure is likely to increase due to changes in economic development, which increases the value and density of human and physical capital (EEA 2007a).

Storms are currently the costliest weather catastrophes in the developed world and they are likely to become more powerful in the future as the oceans warm and provide more energy to fuel storms (Stern 2006). Cost effects will be magnified for extreme storms, which are expected to disproportionately increase costs in relation to an average storm. ABI estimated that wind-related insurance losses from extreme European storms will increase by at least 5% to EUR 25–30 billion. Some estimates indicate that the cumulative contribution of changing climate and socioeconomic development are likely to double worldwide economic losses due to natural disasters every 10 years.

Some preliminary estimates (ABI 2005) indicate that annual flood losses in Europe could rise to EUR 100–120 billion (10-fold) by the end of the century (though flood management could reduce this). Besides projected growth in direct damage to settlements and infrastructure, an increase in the number and severity of flood events will affect large parts of European industry and power generation located in flood prone areas. This may create competitiveness concerns due to an increased risk of business interruption.

Looking at other events, the total loss of the hot summer of 2003 in France, including power generation, stress on the transport system, stress on forests and other ecosystems including fires, reduced wine production and decreased agricultural productivity, has been estimated at 0.1–0.2% of the GDP, equivalent to 15–30 billion euros. The summer of 2003 is also estimated to have increased claims by 20% in the United Kingdom, with estimated impacts of GBP 30–120 million and damage to transport infrastructure (rail buckling and road damage) of GBP 40 million (EEA 2007a). Other studies also predict likely future trends in other parts of Europe and show potentially strong increases in future economic costs.

Adaptation will be able reduce costs and disruption caused by extreme weather events such as storms, floods and heat waves, however at higher temperatures, the cost of adaptation will rise sharply and residual damages will remain large. The additional costs of creating new infrastructure and buildings more resilient to climate change in OECD countries could range from 0.05 to 0.5% of GDP each year, with costs likely rising parallel to the higher temperatures expected in the future

(Stern 2006). It is likely that increased claims due to losses induced by more frequent and intense climate change events or increased difficulty in their prediction will probably translate into increases in risk premiums, and/or increases in the levels of uninsured assets.

3.3.5 *Human Health*

The big environmental issues threatening human health in the world today are climate change, ozone depletion, ecological degradation and biodiversity loss, water degradation, acid deposition, local air pollution and land degradation. Climate change poses significant risks to the right to good health. The World Health Organisation states that global warming may already be responsible for more than 160,000 deaths a year from malaria and malnutrition; a number that could double by 2020 (WHO 2003).

Climate change will impact human health in many ways. It will affect the intensity of a wide range of diseases – vector-borne, water-borne and respiratory. In the Pacific, changes in temperature and rainfall will make it harder to control dengue fever. In Australia, there is a risk that the range and spread of tropical diseases and pests will increase. A warmer climate will provide a more hospitable environment for disease-carrying mosquitoes. Climate change is already having a tangible effect on human health. The spectrum of illnesses and diseases is changing in Europe too; allergies increased and infectious diseases spread. Heat waves place a great strain on older people with cardiovascular diseases. Furthermore, climate change is likely to facilitate the spread of some infectious diseases, since the living conditions of animals carrying such diseases are improving. Climate change will place an additional burden on health systems, especially in EU regions that already suffer from shortages and poor supply in the health sector. Alongside heat and infectious disease warning systems already in place, it is essential to raise public awareness of existing dangers and prepare those who will be affected for the impacts of climate change as fully as possible in order to ensure the best possible health protection for the population. As climate change is a global problem it is essential to have a global cooperation network with a view to sustainable adaptation of health systems.

We can conclude that for most sectors, distribution of effects will be strongly geographical (spatial) across Europe, including economic effects. While there is a range of positive and negative economic effects across sectors and regions, there appears to be a significant trend towards more negative impacts in Southeastern Europe and the Mediterranean, especially in relation to energy demands, agricultural productivity, water availability, health effects, summer tourism and ecosystems. Future trends in economic, social, institutional, and technological development in these regions also needs to be taken into account. Many impacts can be avoided, reduced or delayed through mitigation, but adaptation is also necessary. It is very likely that anthropogenic climate change will result in net damage costs into the future and can impede nations' abilities to achieve sustainable development pathways.

3.4 Conclusions

Undoubtedly humanity is already managing the planet, but in an unconnected and haphazard way driven ultimately by individual and group needs and desires. As a result of the innumerable human activities that perturb and transform the global environment, the entire planet is being pushed beyond its natural operating domain. Many of these global changes are accelerating due to a rapidly growing world population.

Our common future is at risk from climate change. Behind the summary numbers lies a story of multiple, interacting, and worsening harms. A 2°C change will be bad enough; changes beyond that point become increasingly perilous. Even in the short run, the impact of climate change will be disproportionately bad for countries which have had the least to do with adding carbon to the atmosphere. It is vital to create institutions to address equity and provide relief as quickly as possible, before food and water problems in the developing world reach critical levels. Waiting to see if things really turn out as badly as predicted will mean we miss the last chance to ensure that future generations inherit a liveable world.

Europe's climate will continue to change over the coming years with impacts on EU economy and society ranging from water, agriculture, forestry, industry and biodiversity to urban life. Current scenarios and impact studies use relatively crude spatial and temporal resolutions. Despite a growing number of country-level case studies, current knowledge of impacts is still incomplete and does not allow for a careful, detailed comparison across regions. Differences in assumptions often make it difficult to compare studies. Only a few studies provide a consistent picture, based on uniform assumptions of climate, socioeconomics, etc. and many studies extrapolate between regions. There is a need for consistent European studies.

Therefore, in addition to avoiding and reversing climate change through reducing emissions of greenhouse gases, there is an urgent need to ensure that we are able to adequately adapt to climate change, which is complex. It involves considering impacts on a range of sectors, organizations and people. Major advances are needed to understand the economics of adaptation to global change, as well. Adaptation will entail complex behavioural, technological and institutional adjustments at all levels of society, and not all population groups will be equally capable of adapting. Such analysis is complicated by the strong link between adaptation and socioeconomic scenarios and development.

The impacts of climatic changes will hit local and regional levels in different ways. The majority of adaptive actions therefore need to be decided and undertaken at local, regional and national levels. Priorities are to integrate climate change impacts and adaptation into sectoral planning, to improve capacities to address extreme weather events and include climate change aspects into long-term investments. Adaptation needs the participation of all stakeholders involved in policy, business or service which are or will be affected by climate change. The scientific community should help in this process by mitigating the misconception that adaptation strategies and subsequent actions are always expensive to implement and that non-action is a cheaper alternative.

References

ABI (2005) Financial Risks of Climate Change, http://www.abi.org.uk/Display/File/Child/552/Financial_Risks_of_Climate_Change.pdf

Alexandrov V, Eitzinger J, Caijc V, Oberfoster M (2002) Potential climate change impacts on selected agricultural crops in northeast Austria. Global Change Biol 8(4):372–389.

Audsley E, Pearn KR, Simota C, Cojocaru G, Koutsidou E, Rounsevell MDA, Trnka M, Alexandrov V (2006) What can scenario modelling tell us about future European scale land use, and what not? Environ Sci Policy 9(2):148–162.

Bindi M, Moriondo M (2005) Impact of a 2°C global temperature rise on the Mediterranean region: Agriculture analysis assessment. In: Giannakopoulos C, Bindi M, Moriondo M, Tin T (eds), Climate change impacts in the Mediterranean resulting from a 2°C global temperature rise, WWF, pp. 54–66.

COPA COGECA (2003) Factsheets: Assessment of the impact of the heat wave and drought of the summer 2003 on agriculture and forestry http://www.copa-cogeca.be/en/dossiers.asp

EEA (2004a) Impacts of Europe's changing climate: an indicator-based assessment, EEA Report No 2/2004, 107 pp.

EEA (2004b) EEA Signals 2004. A European Environmental Agency update on selected issues. 31 pp.

EEA (2005a) European Environmental Outlook. EEA Report No. 4/2005, 85 pp.

EEA (2005b) Vulnerability and adaptation to climate change in Europe. EEA Technical report No 7/2005, 79 pp.

EEA (2007a) Climate change: the cost of inaction and the cost of adaptation. EEA Technical report No 13/2007.

EEA (2007b) Climate change and water adaptation issues. EEA Technical report. No 2/2007.

European Commission (2007) Green Paper from the Commission to the Council, the European Parliament, the European Economic and Social Committee and the Committee of the Regions – Adapting to climate change in Europe – options for EU action {SEC(2007) 849}.

Giorgi F, Bi XQ, Pal J (2004) Mean, interannual variability and trends in a regional climate change experiment over Europe. II: Climate change scenarios (2071–2100). Clim Dyn 23(7–8):839–858.

Goodess C (2005) STARDEX – Downscaling Climate Extremes, UEA, Norwich.

IPCC (2001) Climate Change 2001: The Scientific Basis, Intergovernmental Panel on Climate Change, http://www.grida.no/climate/ipcc_tar/wg1/index.htm

IPCC (2007a) Climate Change 2007: The Physical Science Basis. Contribution of Working Contribution of Working Group I to the Fourth Assessment Report of the Intergovernmental Panel on Climate Change (Solomon S, Qin D, Manning M, Chen Z, Marquis M, Averyt KB, Tignor M, Miller HL, eds). Cambridge University Press, Cambridge, United Kingdom and New York, NY, USA.

IPCC (2007b) Climate Change 2007: Mitigation. Contribution of Working Group III to the Fourth Assessment Report of the Intergovernmental Panel on Climate Change (Metz B, Davidson OR, Bosch PR, Dave R, Meyer LA, eds), Cambridge University Press, Cambridge, United Kingdom and New York, NY, USA, 851 pp.

IPCC (2007c) IPCC Fourth Assessment Report Summary for Policymakers of the AR4 Synthesis Report, IPCC, Geneva.

Maracchi G, Sirotenko O, Bindi M (2004) Impacts of present and future climate variability on agriculture and forestry in the temperate regions: Europe. Clim Change 70(1): 117–135.

Metzger MJ, Leemans R, Schröter D, Cramer W, the ATEAM consortium (2004) The ATEAM Vulnerability Mapping Tool. Quantitative Approaches in System Analysis No. 27. Wageningen, C.T. de Witt Graduate School for Production Ecology and Resource Conservation, Wageningen, CD ROM.

Olesen JE, Carter TR, Díaz-Ambrona CH, et al. (2007) Uncertainties in projected impacts of climate change on European agriculture and terrestrial ecosystems based on scenarios from regional climate models. Clim Change 81:123–143.

PESETA (2007) http://peseta.jrc.es/docs/Agriculture.html

PRUDENCE (2005) Prediction of Regional scenarios and Uncertainties for Defining EuropeaN Climate change risks and Effects. Final Report (http://prudence.dmi.dk)

Parry M (ed) (2000) Assessment of Potential Effects and Adaptations for Climate Change in Europe: The Europe ACACIA Project, Jackon Enviornment Institute, University of East Anglia, Norwich, 320 pp.

Richter G, Semenov M (2005) Re-Assessing Drought Risks for UK Crops using UKCIP02 Climate Change Scenarios. Final report of DEFRA Project CC0368.

Räisänen J, Hansson U, Ullerstig A, Doscher R, Graham LP, Jones C, Meier HEM, Samuelsson P, Willen U (2004) European climate in the late twenty-first century: regional simulations with two driving global models and two forcing scenarios. Clim Dyn 22(1):13–31.

Schröter D, Cramer W, Leemans R, Prentice IC, Araujo MB, Arnell NW, Bondeau A, Bugmann H, Carter TR, Garcia CA et al. (2005) Ecosystem service supply and human vulnerability to global change in Europe. Science 310:1333–1337.

Stern N (2006) The Economics of Climate Change. The Stern Review. Cambridge University Press, available for download at http://www.hm-treasury.gov.uk/index.cfm

WHO (2003) Climate Change and Human Health – Risks and Responses – Summary, WHO, Geneva, pp 37.

Chapter 4
Science, Society and Action

Marieta P. Staneva

Abstract This chapter traces the history of the National Coordination Center for Global Change in the Bulgarian Academy of Sciences (NCCGC, now the Scientific Coordination Center for Global Change, SCCGC), highlighting its decade of existence. It presents a brief overview of the ensuing research projects, studies, books, publications and educational initiatives, which are the result of intensive and fruitful collaboration in the realm of climate change and sustainable development among the Pennsylvania State University, the US National Science Foundation (NSF), institutes of the Bulgarian Academy of Sciences (BAS) and Bulgarian universities. Significant achievements of this collaboration were establishment of the NCCGC by BAS in 1997 as well as the books "Global Change and Bulgaria", and "Drought in Bulgaria: A Contemporary Analog of Climate Change". Other outcomes are computer-based climate and hydro-economic models in a geographic information systems (GIS) framework for the Struma and Yantra River basins, which would assist decision makers with planning in Bulgaria to reduce negative impacts and vulnerability of the country in the face of climate change. Research results and lessons learned are viewed as road markers to assist decision makers and the public in the face of climate change and planning for sustainability. The chapter urges further contributions of the SCCGC to Bulgarian climate challenges.

Keywords Drought · Climate change · Bulgarian society · Decision makers

4.1 Introduction Through a Retrospective Lens

The conference leading to this book was initiated to mark one decade since the founding of the National Coordination Center for Global Change in the Bulgarian Academy of Sciences. Looking back helps us realize how far Bulgaria has come.

M.P. Staneva (✉)
Department of Geography, Pennsylvania State University, University Park, PA, USA
e-mail: mps5@psu.edu

V. Alexandrov et al. (eds.), *Global Environmental Change: Challenges to Science and Society in Southeastern Europe*, DOI 10.1007/978-90-481-8695-2_4,

Coincidentally the years of 2007–2009 marked other important anniversaries. Two decades have passed since the first civil protest in the name of protecting citizen health in Ruse city that was being endangered by cross-boundary pollution of chlorine leaks along the Bulgarian-Romanian border on the Danube River. A commemorative exhibit was organized in September of 2007, revealing known and previously unknown facts about the courage of the people of Ruse, especially its women and mothers of young infants, and their leading organizers and spokesmen, who were prominent individuals: a woman sport coach, a writer, an artist, a historian, a poet, and a journalist among others (Obshtestvonet 2009). One cannot help but notice that scientists were missing from this important scene of action among the rest of the intellectuals. It was not easy to stand up against the then communist authorities and to demand clean air for the city of Ruse – for its children and adults alike. Women from the landscaping department of the city and mothers with their infants in strollers were in the front lines during the demonstration. Some archives displayed in the exhibit twenty years later revealed that the director of the Ruse branch of the Inspectorate for Hygiene and Epidemiology wrote a memo to the higher authorities pointing out that the amount of chlorine measured in the air above Ruse on 23–24 May 1982 was four to five times the then permissible level. It is painful to realize how many years it took the citizens of Ruse to collect their courage, go out in the street and demand responsibility from the local politicians during those difficult and frightening times. It was only in the years of "glasnost" and "perestroika" that the suppressed and dormant forces in the society could be unleashed at last.

A year later – in the autumn of 1989 – another big environmental protest shook the foundation of the established regime. An international environmental event was taking place in the capital city of Sofia, and during those days the citizens organized a civil protest against a megalomaniac hydro-derivation in the Rila Mountains on 26 October 1989 (Omda 2008). The semi-legal movement Ecoglasnost was at the center of the event with one of the leading figures being Petar Beron, an ecologist from the Bulgarian Academy of Sciences. Those beginnings of public expression were swept with enthusiasm and euphoria, as well as naïveté, as noted in a commemorative essay by leading organizers and participants two decades later. The nostalgia is tinged with deep disappointment comparing how much courage was to be had then, and how soon those same people split along political lines, and how the public did not venture enough during the time of transition to stop expansion of the very same project and other construction detrimental to the environment (Bakalov 2008). As we all know, political changes swept the country shortly after the above event, and life became heavily politicized after 10 November 1989. Among mushrooming parties and shifting governments, the life for the ordinary citizen was often unpredictably impacted by economic ups and downs, when survival and making ends meet was the norm of the day. Those processes are well captured and analyzed by Cellarius (2004) and Sardamov (2007), the latter looking at the disillusionment of the Bulgarian society in the past and very recent years.

4.2 Why Bulgaria and Why Research Amidst Economic Turbulence?

It was during those hard times of the beginnings of transition that Penn State faculty members Knight and Staneva spent a year in Bulgaria 1994–1995 under aegis of a Fulbright visit to the new American University in Bulgaria. Their research interest focused on climate change and water resources, which was additionally stimulated by the events surrounding the severe water crisis in the country during the fall and winter of 1994. The core of their research interest and the societal upheaval leading to the "water war" (Staddon 2004) encouraged research for more knowledge about the evolving drought, water management, society and water culture, and related policy-making in Bulgaria. They visited many regional Water and Sewer companies (ViK) trying to understand the management side at the local level and the need for applicable research for decision makers. Frequent interactions with the National Water Council – then the highest governing body for water – brought interesting understanding of the whole dynamics of water issues in the country which was undergoing severe economic, political and drought stress. Another direction of Knight and Staneva's activities was devoted to contacting researchers from various institutes of the Bulgarian Academy of Sciences and universities in the attempt to build bridges among them and prepare grounds for an extensive work on global change. All this was happening while the water crisis was evolving, with contradictions and various viewpoints occupying media's attention (Fotev 2004a, b). Knight and Staneva visited the Rila Mountain site where the people from Sapareva Banya were trying to stand up against the government and powerful construction firms, demanding that local interests should not be ignored for the sake of diverting water to the capital city of Sofia.

The first immediate result from the research interest in water resources was a review of the water situation in Bulgaria (Knight et al. 1995) followed by a special issue of GeoJournal (Knight and Staneva 1996a, b). Another important part of their work was the application of Geographic Information Systems (GIS), used for the first time in Bulgarian research – the Struma river GIS-based model for water resources and climate change scenario (Knight et al. 2001).

In the course of their work in Bulgaria, Knight and Staneva met multiple times with NGO representatives and government officials, among them the Ministry of Environment and Water and the Ministry of Regional Development and Public Works. In their quest to build climate change awareness and engage stakeholders and decision makers, they talked to members of Parliament and met with the President of the Parliament. They personally took copies of their published books to district and university libraries. It was not difficult to notice the political polarization of the water issues in the city of Sofia at that time. The head of the National Water Council, Prof. Dimitar Mandajiev, was fired; he had called for water rationing in the capital much earlier in time. Although limited water rationing in Sofia households might have been less stressful a year or two earlier, it still might have had unfortunate political outcomes for those in power.

It is easy to look back now and judge that many other activities and research attempts could have been carried out to benefit future scientific knowledge – interviews with governmental officials at the time of the water crisis or surveys studying the public opinion during those harsh waterless times. Thus it was not only a tireless scientific quest for knowledge and understanding of the water and drought situation in Bulgaria, but also the strong civil responsibility of scientists to take a message to present and future decision makers to learn from those lessons surrounding the water crisis.

During 1994–1995, Knight and Staneva had worked closely with Prof. Todor N. Hristov who was then director of the Institute of Water Problems in the Bulgarian Academy of Sciences. A year later CIRA (The Center for Integrated Regional Assessment) at Pennsylvania State University, USA, received approval from NSF that it could subsidize research activities in Bulgaria and a conference on global change. Hristov was invited to be the local coordinator for the conference on global change that would be held in Bulgaria in summer 1997.

Several scientific events and publications are characteristic for the early 1990s. Bulgaria participated in the U.S. Department of Energy's Country Studies Program (USCP), creating an inventory of greenhouse gas emissions and potential mitigation, as well as examining vulnerability as part of a national climate change action plan. It was noticeable that Bulgaria had already marked the beginning of some integration among very few scientists from several fields of study. It was also obvious that intensive integration among institutes and scientists representing a wide range of disciplines was weak and almost missing, a result of the strong division among disciplines at all levels of education and research. It was understandable that it would take special effort to bring scientists from different fields together, so the intent of the 1997 conference was to plant the roots for further cross-institutional cooperation on global change research. The respective authors were encouraged to select one or two co-authors from different institutions whose expertise would extend the scope of the working papers as well as help in building bridges across institutions.

The other goal of the 1997 conference was to identify future research frontiers of mutual interest between CIRA and Bulgarian researchers, as well as within the Bulgarian community and between Bulgarians and other international research groups. In late December 1996 and early January 1997, Knight, Staneva, and Hristov commissioned nine working papers for the conference, inviting senior authors from the Bulgarian community to take leadership in each area. The topics that brought coauthors together represent the main presentations and later on the chapters of the book "Global Change and Bulgaria" (Hristov et al. 1999). Some of them are related to greenhouse gas emissions and mitigation; climate variability and change; hydrology; agriculture and climate change; forestry; economics; demographics and social changes; data and information sources; and politics and institutions. Each writing team also received copies of pre-conference reviews so that appropriate comments could beincluded in the conference discussions. It has to be mentioned again that all these efforts were taking place during a time of severe economic crisis, deep politicization and frequently changing governments. At that time the participation of the broad public was still gathering momentum and slowly becoming visible in the society.

The civil society had just started to evolve and the linkages and communication among NGOs, scientists, lay public and decision makers were obscured by other forces – political polarization with frequent political changes, and everyday pressures of economic survival. A survey done in 1998 points that climate change does not rank among the highest on the list of Bulgarians. The pressing needs of the day were very different, on a family scale (O'Connor et al.1999).

The scientific community, the politicians and the lay public were functioning as separate entities without effective interaction and communication on crucial environmental issues. The scientists seemed to continue acting within disciplinary and institutional boundaries and could not reach effectively to the public in order to bring their scholarly explorations in the light and impact policymaking. Forests have been depleted, rapid and often unregulated development was taking place leading to environmental devastation in vulnerable and fragile areas, the number of vehicles had been increasing – primarily second hand-ones with high levels of emissions – and traffic jams started occurring in many cities.

4.3 Founding of the NCGC After the 1997 Conference

A defining event took place in the summer of 1997 during the global change conference. The theme of the conference was "Bulgarian-American Conference on Global Change and Bulgaria" and it was held 17–19 June 1997 in the campus of the American University in Blagoevgrad, Bulgaria. The integrated efforts of scholars across a wide range of disciplines led to a petition to the leadership of BAN to establish what would become the NCCGC under the leadership of the late Academician Dimitar Mishev. Shortly thereafter, the BAN Council established the National Coordination Center for Global Change in the Bulgarian Academy of Sciences on 2 July 1997. This action surpassed the organizers' greatest expectations of possible outcomes. The more immediate goal of identifying research opportunities had also been realized, resulting in on-going discussions for future research collaboration and support. Research priorities defined collaboratively included four major areas: analysis of greenhouse gas emissions and mitigation; climate change forecasting, especially at the regional level; analysis of impacts of climate change on natural ecosystems; and analysis of the human dimensions of climate change in Bulgaria. The conference summarized climate change research about emissions, climate variability and change, hydrology, agriculture, water resources, economics, demographics, data acquisition, and policymaking. The resultant publication laid the foundation of a future collaborative and intensively integrative project focused on drought.

4.4 Bulgaria in the International Climate Change Community

The 1997 conference called for a number of organizational tasks to be undertaken, including establishing the coordinating council in the Bulgarian Academy of Sciences and developing an agenda for Bulgaria to participate in international activities especially in the global change research community in Central and East Europe and Russia. It was the existence of Bulgarian NCCGC that put the country in a much better position among international climate change scientists and opened doors for broader interactions. Programs and organizations like START (The Global Change SysTem for Analysis, Research and Training) and IHDP (the International Human Dimensions Programme) assisted with Bulgarian participation in important global change activities. The introduction of the present book elaborates on many successful outcomes of the functioning of NCCGC. The NCCGC was renamed by BAS after the death of Academician Mishev in 2003, becoming the Scientific Coordination Center for Global Change.

4.5 Drought Book

The conference of 1997 marked an important undertaking – a seminal project on drought during the 1980s and 1990s exploring impacts on environmental elements, society, economy, and human health in view of the drought as an analog of future climate change, including climate scenarios. Bulgarian scientists enthusiastically embraced Knight's suggestion of this study. The project coordinator on the Bulgarian side was Prof. Ivan Raev, then director of the Forest Research Institute in BAS. It was only two years earlier that the country experienced the peak of a severe drought with a difficult time of extreme water rationing. It was considered that if it were not for the collapse of the economy, the water crisis would have been disastrous. The irrigation systems were no longer in use, industry shut down and water use dropped. Even with a collapsed economy, the city of Sofia could not avoid water rationing because the water in the Iskar reservoir was severely depleted due to different causes such as water mismanagement, drought, and a leaking water supply system, plus various political and economic maneuverings that triggered a deepening water conflict. What could be learned from the dynamics of the drought and its impact that could inform planning for climate change, particularly when climate scenarios pointed to a future normal climate that could be as dry as the historic drought (Knight and Staneva 2003). As a result of the collaborative research, two versions of a book were published: a Bulgarian version published by BAS and launched on 28 November 2003 (Raev and Knight 2003) and "Drought in Bulgaria: A Contemporary Analog for Climate Change" was published in Great Britain (Knight et al. 2004). Designed to provide useful recommendations to policymakers, the book is an example of extensive collaboration across disciplines. It explores a breadth of drought processes and impacts including various natural systems as well as economic and social dimensions. Special attention was paid to climatology and hydrology, water resources and management, forestry, agriculture, societal issues, human health, media and culture, and the challenges of data gathering.

The working hypothesis was structured around climate scenarios with the idea that Bulgaria might experience the extremes of 1993–1994 as a future norm. That is why one of the major contributions of the book is in the recommendations extracted from the lessons learned during the drought and its negative impacts on nature and society – an attempt to cast a bridge to the present and future generations of decisions makers in order to make Bulgaria less vulnerable in the face of climate change (Raev and Knight 2003). What could be done better in the future in order to avoid a replica of the response or inaction at the time? Climate change projections warn about significant vulnerability of the country thus the need for profound response – adequate planning and action. For example, introducing drought resistant crops is imperative (Alexandrov and Slavov 2004) as well as increasing the altitude of the planted coniferous forest (Raev and Rosnev 2004).

The November 2003 launching of the Bulgarian book, accompanied by presentations by the leading authors and coordinators, attracted considerable attention. It took place in the main building of BAS. Governmental officials were present among the scientists, and media did its share with sensational publications. The following day a popular newspaper *24 Hours* reported on the event and included an archival photo featuring the public protest against the infamous Rila diversion project, "Djerman-Skakavitsa" and police's reaction in Sapareva Banya during the water crisis of 1994–1995. The title of the newspaper report was loaded with startling words and messages: drought, rats and hepatitis threaten Bulgaria from 2010. Newspaper *Novinar* was more truthful to the purpose of the event.

4.6 Bulgaria and Climate Change: A Decade of SCCGC

When we look back at the beginning of a decade of climate change scientific activity it is inevitable to point out that Bulgaria has made important strides in the realm of research and collaboration. It has been stated in the introduction of the present book that Bulgaria's scientific climate change community has achieved significant progress in the international arena with the existence of the SCCGC. The government of Bulgaria has also fulfilled its obligation ratifying international climate change documents, scientists were participating in numerous collaborative projects, and the NGOs were becoming more influential. It is evident that the intensity of the work on governmental level has been gathering momentum as part of Bulgaria's effort to become a European Union (EU) member. The years following the country's membership to EU are even greater proof that the forces for cohesion and convergence exerted by the European powers are a call for action coming from outside the country via the supranational organization and its various directives. On the other hand the NGO activities are targeting policy decisions more effectively, energizing greater numbers of citizens and trying to bring a stronger message to the politicians. Certainly many of the NGO activities were part of EU-assisted initiatives as well.

Climate change action days and educational events have been quite significant during recent years. An exhibit "Energy Revolution 2005" elaborating on the topic of climate change organized by Green Peace in collaboration with NGOs toured several big cities in the country (Blue Link 2005). It is striking to see an exhibit that is vastly informative, global in scale but skillfully targeting participation and action on individual level. Yet this laudable effort lacked regional content, which could have made it much more focused and effective. How easy it could have been to use illustrations from the book on drought under different climate change scenarios for Bulgaria (Steuer and Knight 2004). An addition to the exhibit bearing a local focus could have been done within a broader regional context – bringing together information from scientific studies done by researchers from Bulgaria and from the Balkans. If the scientific and NGO community failed to do its share by preparing appropriate regional contents to be added to the Green Peace climate change exhibit, then a similar exhibit could be planned for the future under the auspices of SCCGC. This would be an appropriate idea for productive collaboration of Balkan scientists in the climate change community.

Another similar lost opportunity for effective impact and source for collaboration occurred during the chain of interesting activities marking the climate change action day in several recent years. The Climate Action Day in 2007, for example, represented a long list of very informative and engaging initiatives (Blue Link 2007). How unfortunate that such an impressive event aiming at fulfilling a wonderful agenda did not seem to involve the scientific community – it is not mentioned among the supporting organizers. It is as if the scientists are not part of the action, and the regional focus is also missing, which could have been a major contribution for bringing the message to the lay public about the vulnerability of Bulgaria in the face of climate change. One should ask the same question over and over again – where were the scientists, to what does local research contribute, who does it benefit, why is coordination of efforts and synergy missing?

Some of the recent years and especially the summer of 2005 have brought extreme natural phenomena to the country, and the media turned its attention again to scientists and scientific sources. Regretfully, in such an extensive material and interview taken by Nikolov (2005) there is no mention of SCCGC as a Bulgarian focal point of climate change research and communication to the public and politicians.

The imperative need for coordination is exhibited by the fact that another conference with international participation and similar themes was held shortly after the SCCGC conference during 5–6 June 2008 at the University of Architecture, Civil Engineering and Geodesy with many scientific, education and business organizers: Sofia University – Biology Department, Forestry Institute at BAS, Technical University, Bulgarian Chamber of Commerce, National Tourist Board to name a few (Youth Informational Consulting Center 2008). This conference "Ecology and Environment: Regional and National Problems and Trends" was held under the auspices of the President of Bulgaria and its closing session voted a memorandum to the legislative and executive authorities.

If we look at the broader recent picture of Bulgarian society we see that governmental and non-governmental interests and projects show a new direction towards alternative energy, and the stride to energy efficiency is a continuing effort (Blue Link 2008; Za Zemiata 2008). A concern about drought and water resources, and agriculture under changing climate is obvious as well (Bulgaria 2008b; Ecomedia 2009). In January 2009, the Ministry of Environment and Water closed a competition for projects "Development of River Basin Management Plans" under the Operational Program "Environment 2007–2013" (Bulgaria 2008a). It would be regrettable if research results and developed hydro-economic models and management tools from Struma and Yantra River in the framework of GIS were not utilized. The GIS based Struma river hydrological model, mentioned earlier, was expanded upon in the project WATERMAN to include trans-boundary management of water resources (an EU-Copernicus program with participation from Bulgaria, Germany, Macedonia, Greece, Italy and USA). Another collaborative, multilayered river project developed a complex special hydrological-economic decision-support system, REKA (River Environmental Knowledge and Assessment). The system has three components BISTRA (Basin Impacts Simulated Transport from Rural Areas), VODA (Validation and Optimization for Decision Analysis) and PLAN (Protection Location and Action Network). Scenarios could be generated under different economic, climate, hydrological, pollution and other conditions, and choices for making decisions could be assessed by stakeholders for best management of water resources (Knight et al. 2002; Carmichael et al. 2002; Hamlett et al. 2001).

A very recent insightful effort for communication with future politicians and members of Parliament was undertaken by the environmental organization "Za Zemiata" in collaboration with other NGOs in early July of 2009 (Velichkova 2009). Questionnaires were handed to those running for Parliament targeting their stance on energy, climate change and other environmental problems. Most of the answers show nuanced concerns about the nuclear power station and the need for alternative energy sources. There do not seem to be statements by those future members of Parliament and government about their understanding of climate change or its impacts on local and regional scale. Was there input by scientists who have worked on climate change locally, and why the local or regional perspective had not been brought to the attention of the eventual future politicians? A questionnaire could easily become an educational tool, and an important occasion for effectively impacting policy makers may have been overlooked. It would have been appropriate to use some of the recommendations of the book on drought and ask the political candidates for their opinion and readiness for action under a scenario similar to the drought years of 1982–1994.

Are institutions and lay public alike considering scientific guidance in their day-to-day or long term planning in any sector of the economy and society? Missed opportunities are important indicators for future action which could be taken by SCCGC – preparing a pamphlet about the regional climate change scenarios for the policy makers and lay public, round table discussions with politicians on climate change scenarios, training sessions and similar activities beyond basic and applied scientific research. Looking a decade back and ahead into the future one could pause to ask the question: Does the science-politicians-society dialogue remain a Gordian knot in Bulgaria? Can the SCCGC do more?

References

Alexandrov V, Slavov N (2004) Drought impacts on crops. In: CG Knight, I Raev and MP Staneva (eds) Drought in Bulgaria: A Contemporary Analog of Climate Change, Ashgate Press, Aldershot, UK, pp. 157–170.

Bakalov I (2008) Hydro Diversion Protest of 1989. http://e-vestnik.bg/4944 (accessed 10/24/2009)

Blue Link (2005) Green Peace Exhibit "Energy Revolution: 2005". http://bluelink.net/climate/nfp.shtml?x=7194 (accessed 10/24/2009)

Blue Link (2007) Climate Action Bulgaria – 2007. http://www.bluelink.net/climate/nfp.shtml?x=26263 (accessed 10/24/2009)

Blue Link (2008) Alternative energy in Bulgaria: Solar and wind. http://www.bluelink.net/climate/nfp.shtml?x=37859 (accessed 10/24/2009)

Bulgaria. Ministry of Environment and Water (2008a) Development of River Basin Management Plans under Operational Programme Environment 2007–2013. http://ope.moew.government.bg/en/notice/noticedetail/from/noticearchive/id/10 (accessed 10/24/2009)

Bulgaria. Ministry of Environment and Water (2008b) Program on water under drought. http://www.moew.government.bg/recent_doc/strateg_plans/Netty10.doc (accessed 10/24/2009)

Carmichael JD, Dimitrov C, Knight G (2002) Water quality and economic analysis – the VODA Model. In: CG Knight and TN Hristov (eds) River Environmental Knowledge and Assessment – REKA, Applied to the Yantra River Basin, National Coordination Center for Global Change, Bulgarian Academy of Sciences, Sofia, pp. 57–75.

Cellarius B (2004) The fieldwork context: A brief history of postsocialist Bulgaria. In: B Cellarius (ed) In the Land of Orpheus: Rural Livelihoods and Nature Conservation in Postsocialist Bulgaria, The University of Wisconsin Press, Wisconsin, MI, pp. 15–27.

Ecomedia (2009) Climate change and agriculture. http://ecomedia.bg/news/bulgaria-news/article/353 (accessed 10/24/2009)

Fotev G (2004a) Bulgarian society and water crisis: Sociology and ethics. In: CG Knight, I Raev, and MP Staneva (eds.) Drought in Bulgaria: A Contemporary Analog of Climate Change, Ashgate Press, Aldershot, UK, pp. 185–202.

Fotev G (2004b) Media and political attitudes. In: CG Knight, I Raev, and MP Staneva (eds) Drought in Bulgaria: A Contemporary Analog of Climate Change, Ashgate Press, Aldershot, UK, pp. 277–288.

Hamlett JM, Knight CG, Evans BM, Carmichael JJ, Hristov TN, Ioncheva VD, Dimitrov D, Nikolov II, Chang H, Staneva MP (2001) Yantra Basin Decision Support System: Linking GIS with nonpoint source and in-stream models. The American Society of Agricultural Engineers (ASAE), Microfiche No. 01-2123, ASAE, St. Joseph, MI.

Hristov T, Knight CG, Mishev D, Staneva M (eds) (1999) Globalnite promeni I Bulgaria: Vuvedenie (Global Change and Bulgaria: Introduction). In: Globalnite Promeni i Bulgaria (Global Change and Bulgaria), National Coordination Center for Global Change, Bulgarian Academy of Sciences, Sofia, pp. 5–16.

Knight CG, Velev SP, Staneva MP (1995) The emerging water crisis in Bulgaria. GeoJournal 35(4):415–423.

Knight CG, Staneva MP (1996a) Water resources of Bulgaria: Introduction to a special issue of GeoJournal. GeoJournal 40(4):343–345.

Knight CG, Staneva M (1996b) Water resources of Bulgaria: An overview. GeoJournal 40(4):347–362.

Knight CG, Heejun C, Deyan K, Marieta PS (2001) A simplified basin model for simulating runoff: The Struma River GIS. Prof Geo 53(4):533–545.

Knight CG, Hristov TN, Staneva MP (2002) Introduction. In: River Environmental Knowledge and Assessment – REKA, Applied to the Yantra River Basin, National Coordination Center for Global Change, Bulgarian Academy of Sciences, Sofia, pp. 1–4.

Knight CG, Staneva MP (2003) Аналозите в изследванията за промяна на климата. In: I Raev, CG Knight, MP Staneva (eds) Засушаването в България: съвременен аналог за климатични промени, Bulgarian Academy of Sciences Press, Sofia, pp. 30–37, (in Bulgarian).

Knight CG, Raev I, Staneva MP (2003) Въведение. In: I Raev, CG Knight, MP Staneva (eds) Засушаването в България: съвременен аналог за климатични промени, Bulgarian Academy of Sciences Press, Sofia, pp. 13–19, (in Bulgarian).

Knight CG, Raev I, Staneva MP (2004) Drought in Bulgaria: A Contemporary Analog of Climate Change. Ashgate Studies in Environmental Policy and Practice. Aldershot, UK.

Nikolov K (2005) This is not the last nightmarish summer: What will happen to Bulgarian climate in the next 100 years. http://www.mediapool.bg/show/?storyid=109161 (accessed 10/25/2009)

Obshtestvonet (2009) Ruse eco-protest of 1987. http://www.obshtestvo.net/content/view/1400/9 (accessed 10/24/2009)

O'Connor Robert E, Richard J B, Ann F, Marieta S, Kozhouharova-Zhivkova V, Dobreva S (1999) Determinants of support for climate change policies in Bulgaria and the USA. Risk Decis Polic 4(3):255–269.

Omda (2008) Declaration and eco-protest of October 26, 1989. http://www.omda.bg/arhiv/dokumenti_politicheski/26_10_1989.htm (accessed 10/25/2009)

Raev I, Knight CG (2003) Препоръки към вземащите решения. In: I Raev, CG Knight, MP Staneva (eds) Засушаването в България: съвременен аналог за климатични промени, Bulgarian Academy of Sciences Press, Sofia, pp. 274–284, (in Bulgarian).

Raev I, Rosnev B (2004) The impact of drought on natural forest ecosytems. In: CG Knight, I Raev and MP Staneva (eds) Drought in Bulgaria: A Contemporary Analog of Climate Change, Ashgate Press, Aldershot, UK, pp. 117–136.

Sardamov I (2007) From despair to dejection: Bulgaria's transitional pessimism. In: International Conference on "Post-Communist Social and Political Conflicts: Citizenship and Consolidation in the New Democracies of South East Europe," Bucharest, Romania.

Staddon C (2004) The struggle for Djerman-Skakavitsa: Bulgaria's first post-1989 "water war". In: CG Knight, I Raev and MP Staneva (eds) Drought in Bulgaria: A Contemporary Analog of Climate Change, Ashgate Press, Aldershot, UK, pp. 289–306.

Steuer C, Knight CG (2004) Drought and future climate. In: CG Knight, I Raev and MP Staneva (eds) Drought in Bulgaria: A Contemporary Analog of Climate Change, Ashgate Press, Aldershot, UK, pp. 19–26.

Velichkova V (2009) Questionnaire to those running for parliament in the elections of July 2009. http://bluelink.net/climate/nfp.shtml?x=41537 (accessed 10/24/2009)

Youth Informational Consulting Center (2008) Mladejki Informatsionen Konsultantski Tsentar. Environmental and Climate Change Conference. http://www.sfok.mikc.bg/node/153 (accessed 10/25/2009)

Za Zemiata (2008) Energy efficient street lights. http://www.zazemiata.org/bw/energyeff_bg.php (accessed 10/25/2009)

Part II
Global Change and Climate Change in SEE

Chapter 5
Climate Modelling Beyond the Complexity: Challenges in Model-Building

Dragutin T. Mihailovic

Abstract Among the most interesting and fascinating phenomena (that are predictable) are the complex ocean/atmosphere/land dynamical system called weather and its long time average climate. This complex dynamical system can be described in a simplified manner by the following features: (a) it evolves in time according to a set of rules; (b) the present conditions determine its future; (c) the rules governing them are usually non-linear; and (d) there may be many interacting variables describing it as a whole. It is widely accepted that the climate and the chemical composition on the Earth have been and are maintained at the steady state by the presence of life itself. Therefore, the climate can be considered through modelling processes on the environmental interface that is defined as an interface between two either abiotic or biotic environments which are in relative motion, exchanging energy through biophysical and chemical processes and fluctuating temporally and spatially regardless of its space and time scale. This interface as a complex system is a suitable area for occurrence of the chaotic irregularities in temporal variation of some physical or biological quantities describing their interaction. In this paper we consider some issues which are important from the point of view of the current climate modelling attempts. We deal with the following points: (i) overview of the results achieved until the present time, (ii) discussion of the term predictability beyond the complexity and (iii) numerical investigation of the system of two coupled logistic representing energy exchange of two interacting environmental interfaces relevant for providing insight into the properties of the Earth's climate.

Keywords Climate modelling · Predictability · Complexity · Gödel's Incompleteness Theorem · Environmental interface energy · Chaos

D.T. Mihailovic (✉)
Faculty of Agriculture, University of Novi Sad, Novi Sad, Serbia
e-mail: guto@polj.ns.ac.yu

V. Alexandrov et al. (eds.), *Global Environmental Change: Challenges to Science and Society in Southeastern Europe*, DOI 10.1007/978-90-481-8695-2_5,

5.1 Introduction

5.1.1 A Short Overview

Among the most interesting and fascinating phenomena that are predictable are the chaotic ocean/atmosphere/land system called weather and its long time average climate. While weather is not predictable beyond a few days, aspects of the climate may be predictable for years, decades, and perhaps longer (Keller 1999). These two phrases clearly summarise the current opinion and state in climate modelling community that deals with the aforementioned subjects. However, the question of the weather and climate modelling and predictability has been initiated on early 1960s of the twentieth century, which was elaborated in pioneering works by Edward N. Lorenz (1962, 1963a, b, 1964). He was the first person in the scientific world who explicitly pointed out the following points related to the nonlinear dynamics in atmospheric motion: (i) question of prediction and predictability, (ii) importance of understanding the nonlinearity in modelling procedure, (iii) demand for discovery of chaos and (iv) careful consideration of sensitivity of differential equations in modelling system on initial conditions. Subsequent three decades after appearance of these papers, have been characterised by strong interest for predictability of weather and climate on theoretical and practical level. The following topics have been set in the focus: (1) dynamics of error growth; (2) linear and nonlinear systems (normal modes and optimal modes, nonlinear geophysical systems and scale selection in error growth); (3) predictability of systems with many scales; (4) limit of predictability; (5) weather predictability (growth of errors in GCMs based on Lorenz's analysis); (6) predictability from analogs (targeted observations); (7) climate predictability (predictability of time-mean quantities, predictability of the second kind) and potential predictability; (8) seasonal mean predictability and (9) El Niño-Southern Oscillation (ENSO) chaos, predictability of coupled models and decadal modulation of predictability (Krishnamurthy 1993; Hartmann et al. 1995; Lorenz 1969, 1982; Shukla 1981, 1985; Simmons and Hollingsworth 2002; Lorenz and Karry 1998; Shukla et al. 2000; Zwiers et al. 1998; Jin et al. 1994; Kirtman and Schopf 1998). Because the focus of our paper is complexity and predictability in climate modelling, this overview we finish with the comment by Orell (2003): "Prediction problems have been described by Lorenz as falling into two categories. Problems that depend on the initial condition, such as short- to medium-range weather forecasting, are described as predictions of the first kind, while problems that depend on boundary rather than initial conditions, such as, in many cases, the longer-term climatology, are referred to as predictions of the second kind. Both kinds of prediction will be affected by error in the model equations used to approximate the true system (Hunt 1999; Collins 2002)".

5.1.2 Predictability Behind the Complexity

Earth's atmosphere has evolved into a complex system in which life and climate are intricately interwoven. The interface between Earth and atmosphere as a "pulsating biophysical organism" is a complex system itself. The term complex system we use in Rosen's (1991) sense as it explicated in the comment by Collier (2003): "In Rosen's sense a complex system cannot be decomposed non-trivially into a set of part for which it is the logical sum. Rosen's modelling relation requires this. Other notions of modelling would allow complete models of Rosen style complex systems, but the models would have to be what Rosen calls analytic, that is, they would have to be a logical product. Autonomous systems must be complex. Other types of systems may be complex, and some may go in and out of complex phases". Also, we will explain in which sense the term complexity will be used in further text. Usually, that is an ambiguous term, sometimes used (e.g., Rosen 1991) to refer to systems that cannot be modelled precisely in all respects. However, following Arshinov and Fuchs (2003) the term "complexity" has three levels of meaning: (1) there is self-organization and emergence in complex systems (Edmonds 1999), (2) complex systems are not organised centrally, but in a distributed manner; there are many connections between the system's parts (Kauffman 1993; Edmonds 1999), (3) it is difficult to model complex systems and to predict their behaviour even if one knows to a large extent the parts of such systems and the connections between the parts (Heylighen 1997; Edmonds 1999). The complexity of a system depends on the number of its elements and connections between the elements (the system's structure). According to this assumption, Kauffman (1993) defines complexity as the "number of conflicting constraints" in a system, Heylighen (1996) says that complexity can be characterized by a lack of symmetry (symmetry breaking) which means that "no part or aspect of a complex entity can provide sufficient information to actually or statistically predict the properties of the others parts" and Edmonds (1996) defines complexity as "that property of a language expression which makes it difficult to formulate its overall behavior, even when given almost complete information about its atomic components and their inter-relations". Aspects of complexity are things, people, number of elements, number of relations, non-linearity, broken symmetry, non-holonic constraints, hierarchy and emergence (Flood and Carson 1993).

Generally, predictability refers to the degree that a correct forecast of a system's state can be made either qualitatively or quantitatively. For example, while the second law of thermodynamics can tell us about the equilibrium that a system will evolve to, and steady states in dissipative systems can sometimes be predicted, there exists no general rule to predict the time evolution of systems far from equilibrium, etc. chaotic systems, if they do not approach some kind of equilibrium. Their predictability usually deteriorates with time. To quantify predictability, the rate of divergence of system trajectories in phase space can be measured (Kolmogorov-Sinai entropy, Lyapunov exponents).

Lorenz (1984) discussed several issues in the predictability of weather systems. According to him predictability is defined as the degree of accuracy with which it is possible to predict the state of weather system in the near and also the distant future (predictability in Lorenz's sense). In this paper is assumed that weather predictions are made on the basis of imperfect knowledge of a weather system's present and past states. This rather general statement is comprehensively elaborated by Hunt (1999). He described the fundamental assumptions and current methodologies of the two main kinds of environmental forecast (i.e., weather forecast); the first is valid for a limited period of time into the future and over a limited space-time "target", and is largely determined by the initial and preceding state of the environment, such as the weather or pollution levels, up to the time when the forecast is issued and by its state at the edges of the region being considered; the second kind provides statistical information over long periods of time and/or over large space-time targets, so that they only depend on the statistical averages of the initial and "edge" conditions. Environmental forecasts depend on the various ways that models are constructed. These range from those based on the "reductionist" methodology (i.e., the combination of separate, scientifically based, models for the relevant processes) to those based on statistical methodologies, using a mixture of data and scientifically based empirical modelling. For example, limitations of the predictability in the world of atmospheric motions are concisely discussed in paper by James (2002). In this paper is numerically considered the predictability of a forced nonlinear system, proposed by Lorenz, as a compelling heuristic model of the mid-latitude global circulation.

The above insight of the predictability underlined in the context of the "environmental predictability" (primarily linked to the climate change issues), we finish with the question: Can we significantly "improve" the weather/climate predictions comparing to the level they currently reached? The answer can not be strictly elaborated with either yes or no. An optimistic and acceptable attitude, that prefers option yes, is concisely written down by Hunt (1999) as the phrase: "We concluded that philosophical studies of how scientific models develop and of the concept of determinism in science are helpful in considering these complex issues". If we give advantage to the option no then we do not close the door for the first option. It only means that there exists limitation of the modelling attempts on an epistemological level. To show that, we will use the Gödel's Incompleteness Theorem about Number Theory (Gödel 1931). Basically it says that no matter how one tries to formalise a particular part of mathematics, syntactic truth in the formalisation does not coincide with the set of truths about numbers. In other word Gödel's Theorem shows that formalisations are part of mathematics, but not all of mathematics. There are many ways to look and "read" Gödel's Theorem. One exclusive way is offered by Rosen (1985). According to him the first thing to bear in mind is that both Number Theory and any formalisation of it are both systems of entailment. It is the relation between them, or more specifically, the extent to which these schemes of entailment can be brought into congruence, that is of primary interest. The establishment of such congruencies, through the positing of referents in one of them for elements of the other, is the essence of the modelling relation. In a precise sense, this theorem asserts that a formalisation which all entailment is syntactic entailment is too impoverished in entailment to be congruent to Number Theory, no matter how we try to establish such congruence. This kind of situation is termed complexity by Rosen (1977). Namely, in this light, Gödel's Theorem says that Number Theory is more complex than any of its formalisation, or equivalently, that formalisations, governed by syntactic inference alone, are simpler than Number Theory, governed by syntactic inference alone, are simpler than Number Theory. To reach Number Theory from its formalisations, or more generally, to reach a complex system from simpler one, requires some kind of limiting processes.

Let us return to the question we were asking ourselves after we shortly had considered climate modelling (i.e., predictability) beyond the complexity. To our mind there is a significant space for "improvement" of models and their capabilities to provide good forecasts. It can be done only if the modelling attempts are directed towards the following steps: from structures and states to processes and functions; from self-correcting to self-organising systems; from hierarchical steering to participation from conditions of equilibrium to dynamic balances of non equilibrium; from single trajectories to bundles of trajectories; from linear causality to circular causality; from predictability to relative chance; from order and stability to instability, chaos and dynamics; from certainty and determination to a larger degree of risk, ambiguity and uncertainty; from reductionism to emergentism and from being to becoming.

In this paper we first address two issues that, to our mind, are important for providing insight into climate modelling. These are (i) an approach in analysis of the energy balance equation for environmental interface and (ii) chaotic response of environmental interface on forcing by visible and infrared radiation.

5.2 Chaotic Response of Environmental Interface on Forcing by Visible and Infrared Radiation

5.2.1 An Analysis of Energy Balance Equation for Environmental Interface

The complexity of the climate modelling is concisely described by Zeng et al. (1990) on the following way: "It is now widely accepted that Earth is a complex system which consists of the biota and their environment. These two elements of the system are closely coupled: the biota regulates the environment (e.g. climate on a planetary scale) and, in turn, the environment restricts the evolution of the biota and dictates what type of life can exist as a consequence of Darwinian natural selection. Changes in one part will influence the other, being opposed by negative feedback or enhanced by positive feedback, and this may lead to oscillation or chaos in the system. Therefore, in order for climate models to predict the consequences of changes caused by human activities (e.g., the increase of greenhouse effect the biota should be included in the model. On the other hand, the biota and the environment of the Earth are so complex that this system, or even a single aspects of feedback in it, cannot yet be adequately described by simple mathematical equations" (underlined by D.M). In this paper they developed a simple model, referred to as daisy world to illustrate the Gaia concept, in which feedbacks from the biosphere minimise fluctuation in climate conditions. They found that periodic, and even chaotic, states can exist when the parameter controlling the feedback between biota and environmental temperature is changed. This is a good background for the issue we will deal with in this section.

We are considering here the Earth's surface as an "environmental interface", defined as an interface between the two either abiotic or biotic environments which are in a relative motion exchanging energy through biophysical and chemical processes and fluctuating temporally and spatially regardless of its space and time scale (Mihailovic and Balaž, 2007). This definition broadly covers the unavoidable multidisciplinary approach in environmental sciences and also includes the traditional approaches in sciences that are dealing with the environmental space. For this interface, visible radiation provides almost all of the energy received by itself. Some of the radiant energy is reflected back to space. The interface also radiates some of the energy received from the sun. The quantity of the radiant energy remaining on the environmental interface is the net radiation, which drives certain physical processes important to us. Since all of the energy transfer processes occurs in the finite time interval, the energy balance equation at any environmental interface can be written in terms of finite differences of ground and air temperatures and then, under some conditions, further transformed into the logistic equation (Mihailovic et al., 2001).

Our basic equation is the energy balance equation. Since all of the energy transfer processes occur in the finite time interval we shall immediately write this equation in terms of finite differences i.e. in the form of difference equation

$$\mathrm{D}T_i = \mathrm{F}_n \tag{5.1}$$

where D is the finite difference operator defined as $\mathrm{D}T_i = (T_{i,\,n+1} - T_{i,\,n})/\mathrm{D}t$, T_i is the environmental interface temperature, n is the time level, $\mathrm{D}t$ is the time step, $\mathrm{F}_n = (R_n - H_n - E_n - S_n)/c_i$ is defined at the nth time level, R is the net radiation, H, and E are the sensible and the latent heat, respectively, transferred by convection, and S is the heat transferred by conduction into deeper layers of underlying matter while c_i is the environmental interface soil heat capacity per unit area. Equation (5.1) is also written in the finite difference form from an additional reason. It can be explain if we follow comprehensive consideration by van der Vaart (1973) about replacing given differential equations by appropriate difference equations in modelling of phenomena in physical and biological world. According to him, many mathematical models for physical and biological problems have been and will be built in the form of differential equations or systems of such equations. With the advent of computers one can been able to find (approximate) solutions for equations that used to be intractable. Many of the mathematical techniques used in this area amount to replacing the given differential equations by appropriate difference equations, so that extensive research has been done into how to choose appropriate difference equations whose solutions are "good" approximations to the solutions of the given differential equations. In Eq. (5.1) the sensible heat is calculated as $C_H(T_i - T_a)$where C_H is the sensible heat transfer coefficient and $T_a(t)$ is the gas temperature given as the upper boundary condition. The heat transferred into underlying soil matter is calculated as $C_D(T_i - T_d)$ where C_D is the heat conduction coefficient while $T_d(t)$ is the temperature of deeper layer of underlying matter that is given as the lower boundary condition. Following Bhumralkar (1975) and Holtslag and van Ulden (1975) the net radiation term can be represented as $C_R(T_i - T_a)$ where C_R is the radiation coefficient. According to Mihailovic et al. (2001), for small differences of T_a and T_d, the expression for the latent heat can be written in a form$C_L f(T_a[b(T_i - T_a) + b^2(T_i - T_a)^2]/2)$. Here C_L is the latent heat transfer coefficient, $f(T_a)$ is the gas vapour pressure at saturation and b is a constant characteristic for a particular gas. Calculation of time dependent coefficients C_R, C_H, C_L and C_D can be found in Monteith and Unsworth (1990). After collecting all terms in Eq. (5.1) we get

$$\mathrm{D}T_i = A(T_{i,n} - T_{a,n}) - A_2(T_{i,n} - T_{a,n})^2 - A_3(T_{i,n} - T_{d,n}) \qquad (5.2)$$

where $A_1 = [C_R - C_H - C_L bf(T_a)]/c_i$, $A_2 = b^2 f(T_a)/(2c_i)$ are coefficients also depending on $\mathrm{D}t$. With $\mathrm{D}t_p = 1/(A_1 - C_D/c_i)$ we indicate the scaling time range of energy exchange at the environmental interface including coefficients, that express all kind of energy reaching and departing the environmental interface. For any chosen time interval, for solving Eq. (5.2), there always exists $\mathrm{D}t_{p,l} = \min[\mathrm{D}t_p(c_i, C_R, C_H, C_L)]$ when energy at the environmental interface is exchanged in the fastest way by radiation, convection and conduction. If we define dimensionless time $\tau = \mathrm{D}t/\mathrm{D}t_{p,l}$ and use lower boundary condition, $T_{d,n} = T_{a,n} - (c_i/C_D)\mathrm{D}T_a$, then Eq. (5.2) after some transformations takes the form of the logistic equation, i.e.

$$\Gamma_{a+1} = \beta\Gamma_n(1 - \Gamma_n) \qquad (5.3)$$

where the symbols introduced have the following meaning: Γ is the dimensionless quantity (Mihailovic et al., 2001) while logistic parameter $\beta = 1 + \tau$ can take values in the interval $1 < \beta < 4$.

5.2.2 Interacting Environmental Interfaces

Under the aforementioned conditions Eq. (5.3) represents energy exchange at the uniform environmental interface. However, in the nature usually we encounter mixture of two or more environmental interfaces, for example grid-cell surface covered by the different land covers. They will interact by exchanging the energy between them. More generally the model of p coupled maps we introduce takes the form

$$\begin{bmatrix} x_1^{n+1} \\ x_2^{n+1} \\ \cdot \\ x_p^{n+1} \end{bmatrix} = \mathrm{F} \begin{bmatrix} x_1^{n} \\ x_2^{n} \\ \cdot \\ x_p^{n} \end{bmatrix} = \begin{bmatrix} 1-(p-1)\varepsilon_1 & \varepsilon_1 & \cdot\ \cdot\ \cdot & \varepsilon_1 & \varepsilon_1 \\ \varepsilon_2 & 1-(p-1)\varepsilon_2 & \cdot\ \cdot\ \cdot & \varepsilon_2 & \varepsilon_2 \\ \cdot & \cdot & \cdot\ \cdot\ \cdot\ \cdot & \cdot & \cdot \\ \varepsilon_p & \cdot & \cdot\ \cdot\ \cdot & \varepsilon_p & 1-(p-1)\varepsilon_p \end{bmatrix} \cdot \begin{bmatrix} f(x_1^n) \\ f(x_2^n) \\ \cdot \\ f(x_p^n) \end{bmatrix} \tag{5.4}$$

or

$$X_{n+1} = \mathrm{F}(X_n) = \mathrm{A} \cdot (\mathrm{f}(X_n))$$

with

$$X = \begin{bmatrix} x^1 \\ x^2 \\ \cdot \\ x^p \end{bmatrix}, \mathrm{f} = \begin{bmatrix} f(x^1) \\ f(x^2) \\ \cdot \\ f(x^p) \end{bmatrix}$$

and

$$\mathrm{A} = \begin{bmatrix} 1-(p-1)\varepsilon_1 & \varepsilon_1 & \cdot\ \cdot\ \cdot & \varepsilon_1 & \varepsilon_1 \\ \varepsilon_2 & 1-(p-1)\varepsilon_2 & \cdot\ \cdot\ \cdot & \varepsilon_2 & \varepsilon_2 \\ \cdot & \cdot & \cdot\ \cdot\ \cdot\ \cdot & \cdot & \cdot \\ \varepsilon_p & \cdot & \cdot\ \cdot\ \cdot & \varepsilon_p & 1-(p-1)\varepsilon_p \end{bmatrix}$$

where ε_i is the coupling parameter while $i = 1, 2, 3, \ldots, p$.

From the reason of simplicity we consider the case when $p = 2$, $\varepsilon_1 = \varepsilon$ and $\varepsilon_2 = 1-\varepsilon$. Then, using Eq. (5.3), with $f(x_1^{n+1}) = \Gamma_{n+1}^{(1)} = s\beta\Gamma_n^{(1)}(1-\Gamma_n^{(1)})$ and $f(x_2^{n+1}) = \Gamma_{n+1}^{(2)} = \beta\Gamma_n^{(2)}(1-\Gamma_n^{(2)})$, we have the system of two linearly coupled logistic maps having the form

$$\Gamma_{n+1}^{(1)} = (1-\varepsilon)s\beta\Gamma_{n+1}^{(1)}(1-\Gamma_{n+1}^{(1)}) + \varepsilon\beta\Gamma_{n+1}^{(2)}(1-\Gamma_{n+1}^{(2)}) \tag{5.5}$$

$$\Gamma_{n+1}^{(2)} = (1-\varepsilon)\beta\Gamma_{n+1}^{(2)}(1-\Gamma_{n+1}^{(2)}) + \varepsilon s\beta\Gamma_{n+1}^{(1)}(1-\Gamma_{n+1}^{(1)}) \tag{5.6}$$

where $\Gamma^{(1)}$ and $\Gamma^{(2)}$ are physical quantities in the logistic equations (Eq. (5.3)) which are applied on the areas with fractional covers ε and $1-\varepsilon$, respectively, while s is the parameter taking a value between 0 and 1. Note that system (5.5)–(5.6), in the case of $s = 1$, is completely symmetric with respect to exchange of the variables

$\Gamma^{(1)} \leftrightarrow \Gamma^{(2)}$, due to our choice of symmetrically coupled identical subsystems. The interesting is the case of $s \neq 1$. In the following we fix the logistic parameters above and below critical value $\beta = 3.56994$ for β, and $s\beta$, respectively.

The asymptotic behaviour of a series of iterates of the map can be characterised by the largest Lyapunov exponent which, at any initial point $\vec{x}_0$ in an attracting region, is defined by

$$\lambda = \lim_{N\to\infty} \left\{ \ln \left[\left\| \underline{D}^{(N)} (\vec{x}_0) / N \right\| \right] \right\} \tag{5.7}$$

where $\left\| \underline{D}^{(N)} \right\|$ is the norm of the derivative matrix, $\vec{x}_n$ comes from general vector mapping while N is the number of the last iterate. We calculated the Lyapunov exponent λ to see the behaviour of the coupled maps given by Eqs. (5.5) and (5.6) depending on different values of coupling parameter ε. Figure 5.1 shows Lyapunov exponent for the coupled maps as a function of ε ranging from 0 to 1. Each point was obtained by iterating many times from the initial condition to eliminate transient behaviour and then averaging over another 50 000 iterations starting from initial condition $\Gamma^{(1)} = 0.20$ and $\Gamma^{(2)} = 0.25$ with 500 ε values.

This simple analysis, where we consider only Lyapunov exponent, shows a very interesting features of two coupled logistic maps representing interaction of two environmental interfaces through exchange of energy between them. For, example it is seen from Fig. 5.1 that the region with positive Lyapunov exponent, respect to ε, is more emphasised when the one logistic map has lower values of the logistic parameter. Moreover, when the logistic parameters are close each other (violet line), the Lyapunov exponent is always positive expressing the fact that the considered dynamic system is in the chaotic regime for any value of ε.

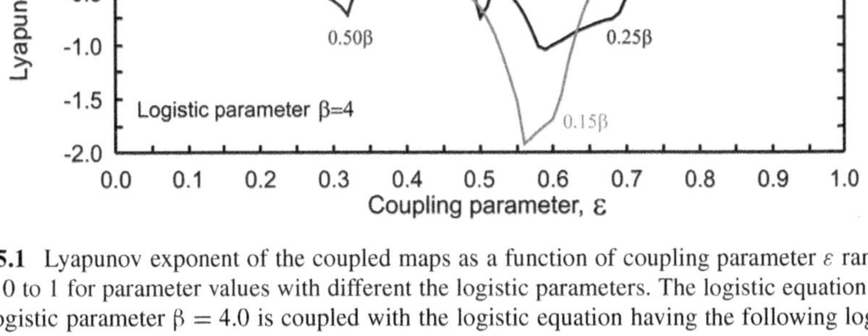

Fig. 5.1 Lyapunov exponent of the coupled maps as a function of coupling parameter ε ranging from 0 to 1 for parameter values with different the logistic parameters. The logistic equation with the logistic parameter $\beta = 4.0$ is coupled with the logistic equation having the following logistic parameters: $s\beta = 0.15\ \beta$; $0.25\ \beta$; $0.50\ \beta$; $0.75\ \beta$ and $0.95\ \beta$

5.3 Conclusion

We considered some issues which are relevant for climate modelling. First, we gave a detailed overview of literature related to this subject. Second, we considered the climate modelling in the light of Gödel's Theorem that says that Number Theory is more complex than any of its formalisation; further we clearly underlined the Rosen's definition of complexity and predictability. Third, we gave an example of analysis of the energy balance equation for environmental surface that provides insight into the properties of the Earth's climate. To illustrate this analysis we performed numerical investigation of the system of two coupled logistic maps, representing energy exchange of two interacting environmental interfaces. It has been done by calculating the Lyapunov exponent as a function of the coupling parameter. It seems that further analysis of this system could be useful providing insight into the properties of the Earth's climate.

Acknowledgement The research work described in this paper has been funded by the Serbian Ministry of Science and Technology under the project "Modelling and Numerical Simulations of Complex Physical Systems", No. ON141035 for 2006-2010.

References

Arshinov V, Fuchs C (2003) Preface. In: Arshinov V and Fuchs C, (eds) Causality, Emergence, Self-Organisation, NIA-Priroda, Moscow, pp. 1–18.

Bhumralkar CM (1975) Numerical experiments on the computation of ground surface temperature in an atmospheric general circulation model. J Appl Meteor 14:1246–1258.

Collier JD (2003) Fundamental properties of self-organisation. In: Arshinov V and Fuchs C, (eds) Causality, Emergence, Self-Organisation, NIA-Priroda, Moscow, pp. 150–166.

Collins M (2002) Climate predictability on interannual to decadal time scales: The initial value problem. Climate Dyn 19:671–692.

Edmonds B (1996) Pragmatic Holism. CPM Report 96-08, MMU.

Edmonds B (1999) What is complexity? The philosophy of complexity per se with application to some examples in evolution. In: Heylighen F and Aerts D (eds) In the Evolution of Complexity, Kluwer, Dordrecht.

Flood RL, Carson ER (1993) Dealing with Complexity: An Introduction to the Theory and Application of Systems Science, Plenum Press, New York.

Gödel K (1931) Über Formal Unentscheidbare Sätze der Principia Mathematica und Verwandter Systeme. I Monatshefte für Math u Physik 38:173–198.

Hartmann DL, Buizza R, Palmer TN (1995) Singular vectors: The effect of spatial scale on linear growth of disturbances. J Atmos Sci 52:3885–3894.

Heylighen F (1996) What is complexity?. In: Heylighen F, Aerts D (eds) In the Evolution of Complexity, Kluwer, Dordrecht.

Heylighen F (1997) The growth of structural and functional complexity. In: Heylighen F, (eds) In the Evolution of Complexity, Kluwer, Dordrecht.

Holtslag AA, van Ulden AP (1975) A simple scheme for daytime estimates of the surface fluxes from routine weather data. J Appl Meteor 22:517–529.

Hunt J (1999) Environmental forecasting and turbulence modeling. Physica D 133:270–295.

Keller DF (1999) Climate, modeling and predictability. Physica D 133:296–308.

James IN (2002) Models of the predictability of a simple nonlinear dynamical system. Atmos Sci Lett 3:42–51.

Jin F-F, Neelin JD, Ghil M (1994) El Niño on the devil's staircase: Annual subharmonic steps to chaos. Science 264:70–72.
Kauffman S (1993) The Origins of Order, Oxford University Press, Oxford.
Krishnamurthy V (1993) A predictability study of Lorenz's 28-variable model as a dynamical system. J Atmos Sci 50:2215–2229.
Kirtman BP, Schopf P (1998) Decadal variability in ENSO predictability and prediction. J Climate 11:2804–2822.
Lorenz EN (1962) The statistical prediction of solutions of dynamic equations. In Proceedings of the Internat. Symp. Numerical Weather Prediction, Tokyo, Japan, November 1960, pp. 629–635.
Lorenz EN (1963a) The predictability of hydrodynamic flows. Trans New York Acad Sci Ser II 25:409–423.
Lorenz EN (1963b) Deterministic nonperiodic flow. J Atmos Sci 20:130–141.
Lorenz EN (1964) The problem of deducing the climate from the governing equations. Tellus 16:1–11.
Lorenz EN (1969) The predictability of a flow which contains many scales of motion. Tellus 21:289–307.
Lorenz EN (1982) Atmospheric predictability experiments with a large numerical model. Tellus 34:505–513.
Lorenz EN (1984) Some aspects of atmospheric predictability. In: Burridge DM, Killn E, eds. In Problems and Prospects in Long and Medium Range Weather Forecasting, Springer-Verlag, Berlin, pp. 1–20.
Lorenz EN, Kerry EA (1998) Optimal sites for supplementary weather observations: Simulation with a small model. J Atmos Sci 55:399–414.
Mihailovic DT, Balaž I (2007) An essay about modelling problems of complex systems in environmental fluid mechanics. Idojaras 111(2–3):209–220.
Mihailovic DT, Kapor DV, Lalic B, Arsenic I (2001) The chaotic time fluctuations of ground surface temperature resulting from energy balance equation for the soil-surface system. In: Abstracts of the 26th General Assembly of European Geophysical Society, Nice, France, 20-25 March 2001.
Monteith JL, Unsworth M (1990) Principles of Environmental Physics, Edward Arnold, London.
Orell D (2003) Model error and predictability over different time scales in the Lorenz'96 systems. J Atmos Sci 60:2219–2228.
Rosen R (1977) Complexity as a system property. Int J Gen Systems 3:227–232.
Rosen R (1985) Anticipatory Systems: Philosophical, Mathematical and Methodological Foundations, Pergamon Press, New York.
Rosen R (1991) Life Itself: A Comprehensive Inquiry into the Nature, Origin and Fabrication of Life, Columbia University Press, New York.
Simmons AJ, Hollingsworth A (2002) Some aspects of the improvement in skill of numerical weather prediction. Quart J Royal Metor Soc 128:647–678.
Shukla J (1981) Dynamical predictability of monthly means. J Atmos Sci 38:2547–2572.
Shukla J (1985) Predictability. Adv Geophys 28B:87–122.
Shukla J, Paolino DA, Straus DM, DeWitt D, Fennessy M, Kinter JL, Marx L, Mo R (2000) Dynamical seasonal predictions with the COLA atmospheric model. Quart J Royal Metor Soc 126B:2265–2291.
van der Vaart HR (1973) A comparative investigation of certain difference equations and related differaential equations: Implications for model building. Bull of Math Biol 35:195–211.
Zeng X, Pielke RA, Eykholt R (1990) Chaos in daisyworld. Tellus 42B:309–318.
Zwiers FW, Kharin VV (1998) Intercomparison of interannual variability and potential predictability: An AMIP diagnostic subproject. Climate Dyn 14:517–528.

Chapter 6
Changes in Different Type of Clouds in South-Eastern Europe in Association with Climate Change

Blanka Bartók

Abstract The aim of the presented study is to estimate regional changes in different cloud types in association with climate change in South Eastern Europe. A statistical method of downscaling is applied to calculate how cloudiness changes parallel to an increase in hemispheric temperature of 0.5°K. The regression of local variables against hemispheric mean temperature is analysed using the method of instrumental variables in the recent monotonously warming period of 1973–1996. The changes in total cloudiness and in different types of clouds (Ci, As, Ac, Cb, Cu, Sc, St and Ns) are both positive and negative, displaying a regional distribution. To find out the geographic distribution of these changes, they are enrolled in clusters and analysed in relation to geographic latitude, longitude and altitude. The results obtained show that total cloudiness, as well as Ac, Cb, Sc and St cloud types have a significant correlation to geographical longitude, while only cirrus clouds changes have a negative correlation to geographical longitude, and only cumulus cloud changes statistically depend on altitude.

Keywords Climate change · Cloudiness · Statistical downscaling

6.1 Introduction

Cloudiness plays a main role in the radiative balance of the atmosphere, thus changes in this component of the climate system require detailed analyses. At the same time, different cloud types have different influences on the climate system, which can lead to modifications of even opposite signs.

Regional changes in cloudiness associated with climate change represent one of the most uncertain aspects of predicting climate change. The complexity of variation

---

B.Bartók (✉)
Faculty of Geography, University of Babeş-Bolyai, Cluj-Napoca, Romania
e-mail: bartokblanka@yahoo.com

V. Alexandrov et al. (eds.), *Global Environmental Change: Challenges to Science and Society in Southeastern Europe*, DOI 10.1007/978-90-481-8695-2_6,

in cloudiness derives, on one hand, from the fact that cloudiness is a local variable associated with anthropogenic activities. On the other hand, the formation of clouds is related to weather situations and geographic factors, such as orography, distance from oceans, etc.

The study presented in this paper aims to analyze changes in the following types of clouds: cirrus, altostratus, altocumulus, cumulonimbus, cumulus, stratocumulus, stratus, nimbostratus (Ci, As, Ac, Cb, Cu, Sc, St and Ns) and total cloudiness in the case of a 0.5°K increase in northern hemispheric temperature. The targeted study area is the South Eastern part of Europe situated between 40°N and 50°N latitude and between 20°E and 30°E longitude (Fig. 6.1).

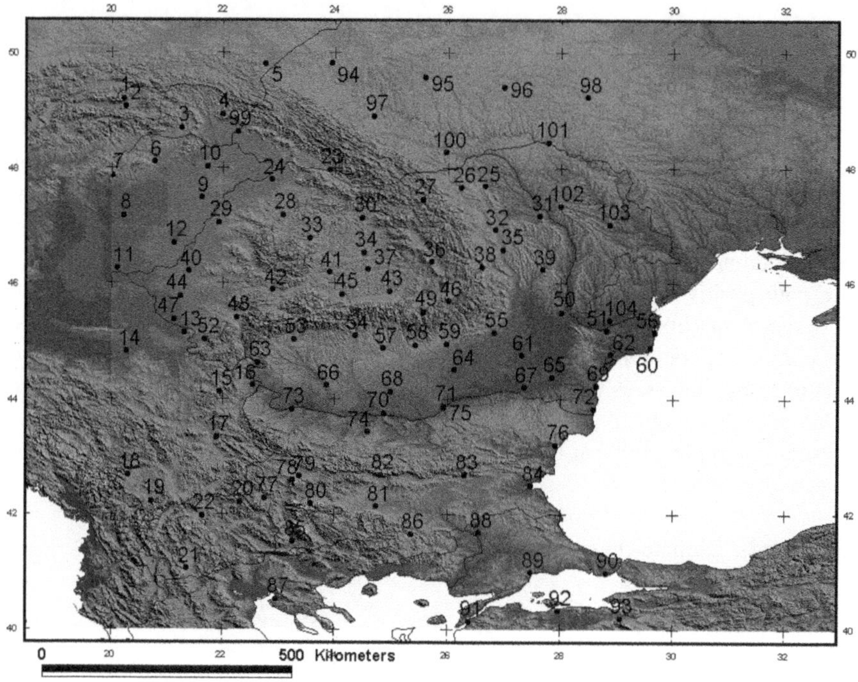

Fig. 6.1 Study area with meteorological stations marked

6.2 Data

Surface-based cloudiness data used in the study are taken from the "Extended Edited Cloud Reports Archive" (Hahn and Warren 1999). The studied region is represented by 104 stations (Fig. 6.1) with statistics taken for the period 1973–1996 as the monthly average amount of total cloud cover is given in percentage of the sky and then separately given for eight cloud types (Ac, As, Ci, Cb, Cu, Ns, Sc and St). Data of northern hemispherical temperature anomalies was reanalysed and published by Jones et al. (Jones 1994).

6.3 Methods

6.3.1 Method of Instrumental Variables

To calculate changes in local cloudiness in parallel with northern hemispheric temperature changes, a statistical downscaling method is applied, namely the method of instrumental variables. Using this method, the regression coefficient is calculated between local variables (cloudiness) and global variables (northern hemispheric temperature anomalies). The criteria for an instrumental variable are:

- non-zero correlation with observed values of the independent variable;
- no correlation to errors of the independent variable;
- no correlation to residuals of regression in the dependent variable.

In case of an instrumental variable, *Z*, linear regression coefficient (*b*) could be estimated as the proportion of appropriate covariance values:

$$b = \frac{\mathrm{Cov}(Y, Z)}{\mathrm{Cov}(X, Z)}$$

For independent variable *X*, hemispheric mean temperature is selected. The instrumental variable for cloudiness is the sequence of years in the 24-year warming-up period (1973–1996) exhibiting high ($r = 0.796$) correlation to hemispheric temperature anomalies and strong (+0.021 K/year) warming trends.

The applied method yields an undistorted point estimation of the regression coefficient, but it is rather difficult to establish significant criteria for these estimations, which is a disadvantage compared to classical regression estimations requiring longer time series. The present approach – developed due to a lack of very long data series – is a combination of that applied by Groisman, Vinnikov and co-workers (overview by Vinnikov 1986) and that suggested by van Loon and Williams (1976). The former uses the application of instrumental variables in estimating regression coefficients of linear connections between hemispheric and regional variables. The latter analyses geographic distributions of local trends in appropriate 30 and 40 year periods from past series when global variables exhibited definite trends.

6.3.2 Cluster Analysis

For analyzing local effects on the behaviour of cloudiness, changes in cloudiness are enrolled in clusters for each type of cloud. The hierarchical cluster method is applied where the number of clusters is five and clusters are determined by the Ward method (Ward 1963) with Euclidean distance using the Statistical Package for the Social Sciences (SPSS) Windows software.

6.4 Results

6.4.1 Changes in Cloudiness Associated with Climate Change

In this section, variation in local cloud cover is calculated in relation to a 0.5°K hemispheric mean temperature increase through the method of instrumental variables for the period 1973–1996. Figure 6.2 shows the behaviour of total cloudiness

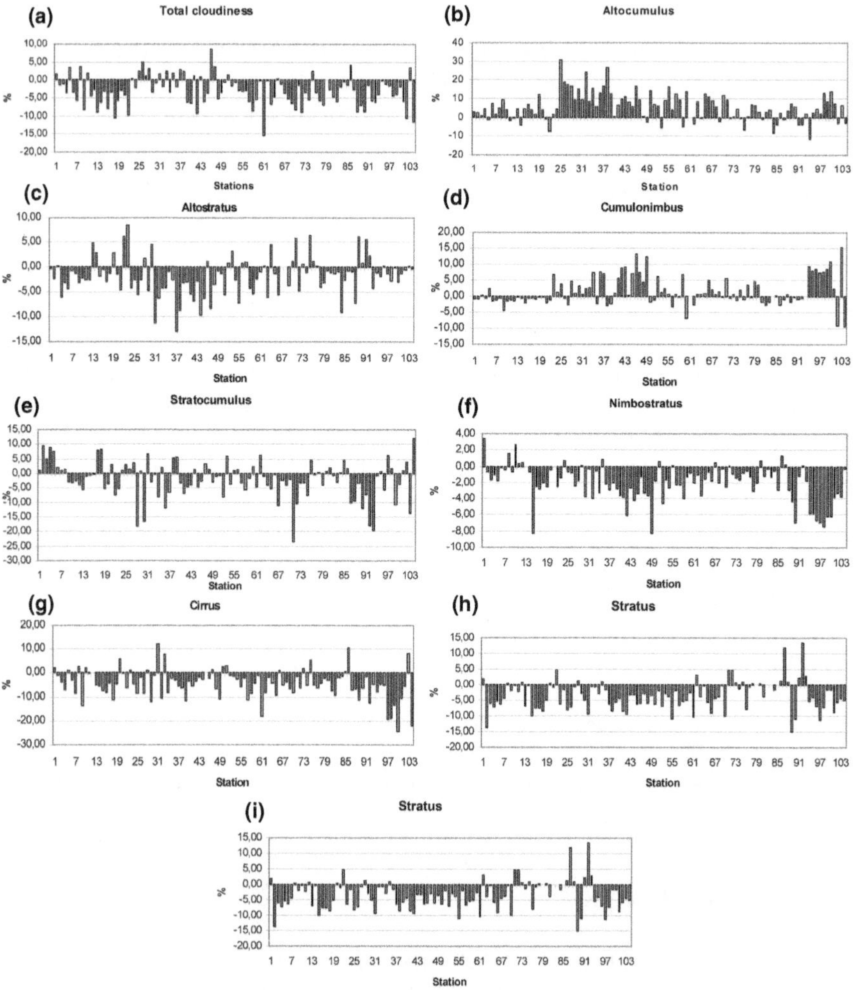

Fig. 6.2 Changes in total cloudiness (**a**) and in different type of clouds (altocumulus-b, altostratus-c, cumulonimbus-d, cumulus-e, stratocumulus-f, nimbostratus-g, cirrus-h and stratus-i) in absolute % in the case of 0.5°K northern hemispheric warming at the 104 meteorological stations which represent the study area

and of the eight types of clouds separately (Ci, Ac, As, St, Sc, Ns, Cu and Cb) in relation to the temperature increase at the 104 stations located within the study area (Fig. 6.1). The given period is characterized by a 0.21 K/decade warming trend. Changes in cloudiness are given in absolute percentiles (100% being equal to an overcast situation). These changes, occurring within a 0.5°K hemispherical mean temperature increase, have the following spatial differences: changes with both positive/increasing and negative/decreasing signs appear within the investigated area.

Total cloudiness follows a mainly negative change (77% of cases) on average by –3.2% absolute percentage (Table 6.1). The maximum negative change is –15.54%. An accentuated decrease is characterised in the case of altostratus, by –1.84% on average (72% of cases), in the case of cirrus by –4.61% (81% of cases), of nimbostratus by –2.03% (85% of cases), of stratocumulus by –2.1% (61% of cases) and of stratus by –3.45% (78% of cases). Only altocumulus exhibited an increase of 5.41%, showing positive change in 76% of all cases. In the case of cumulus and cumulonimbus, average change normally has a small positive value; the proportion of negative and positive changes were around 50%, thus they do not represent an accentuated changing trend on average.

Table 6.1 Average changes (%) of different cloud types associated with climate change in South Eastern Europe

	Ac	As	Ci	Cb	Cu	Ns	Sc	St	Tc
Mean	5.41	−1.84	−4.61	1.57	0.42	−2.03	−2.10	−3.45	−3.02
Max.	30.81	8.32	12.00	15.20	13.22	3.36	12.09	13.46	8.35
Min.	−11.46	−12.96	−24.61	−9.50	−5.46	−8.28	−23.45	−14.91	−15.54
Stdev.	7.37	3.78	6.12	4.38	3.14	2.24	6.35	4.65	4.13
Negative cases %	24	72	81	45	45	85	61	78	77
Positive cases%	76	28	19	55	55	15	39	21	23

6.4.2 Spatial Distribution of Changes in Cloudiness

In Section 6.4.1, absolute changes in different types of clouds are determined through the method of instrumental variables. The results show a spatial distribution of positive and negative changes (Fig. 6.2). A cluster analysis is applied to find out how the different ranges of change are distributed over the study area. Figure 6.3 shows five clusters in the case of total cloudiness changes and further in the case of three types of clouds showing significant correlation to geographical position. Table 6.2 contains average changes in percentage for each cluster, automatically defined by SPSS software. Table 6.3 shows correlation coefficients between the clusters (changing for each type of cloud) as well as for geographical latitude and longitude, and altitude. To calculate correlation between clusters and

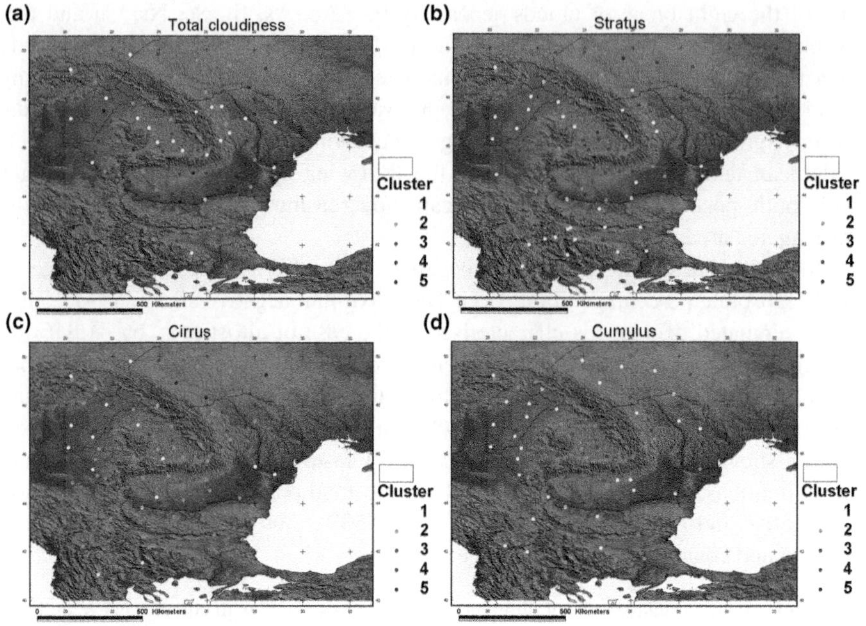

Fig. 6.3 Distribution of the cloudiness changes enrolled in 5 clusters in the case of total cloudiness (**a**) and different types of clouds (stratus-b, cirrus-c, cumulus-d) associated with climate change in the South Eastern part of Europe

other parameters, cluster numbers are reclassified corresponding to maximum and minimum values. For example, cluster number 1 will represent a cluster from the lowest range of values.

Table 6.2 Average values (%) of cloudiness changes for the 5 automatically generated clusters in the case of total cloudiness and for the 8 types of clouds presented in Fig. 4.3

	Clusters				
	1	2	3	4	5
Tc	2.76	−1.05	−3.37	−5.85	−9.73
Ac	1.58	7.77	−5.32	14.43	27.23
As	0.86	−1.94	−5.33	5.75	−10.05
Ci	1.24	−3.25	−8.83	8.23	−20.61
Cb	−0.74	3.73	7.82	12.94	−8.43
Cu	0.71	3.00	−1.21	−4.34	11.44
Ns	3.03	−0.07	−1.58	−6.78	−3.21
Sc	0.55	6.00	−3.72	−9.42	−19.15
St	0.27	−10.91	−6.54	−3.52	12.68

In the case of total cloudiness, only cluster number 1 represents positive change, the other four clusters indicate stations with decreasing cloudiness (Table 6.3). The

Table 6.3 Statistical relationship between cloud clusters and altitude, longitude and latitude (significant correlations marked)

	Ac	As	Ci	Cb	Cu	Ns	Sc	St	Tc
H	0.00	–0.04	0.01	0.07	***–0.21***	0.06	0.16	0.14	0.05
Long.	0.16	0.03	***–0.20***	0.03	0.03	–0.08	–0.13	–0.05	0.06
Lat.	***0.25***	–0.13	–0.01	***0.36***	–0.05	–0.12	***0.26***	***–0.28***	***0.33***

distribution of positive changes depends on geographic latitude (Fig. 6.3a) as the northern part of the area is characterized mainly by positive changes. The statistical relation between the clusters and latitude is significant, with 95% probability as shown on Table 6.3. As a consequence, an increase in total cloudiness can be detected in the northern part and a decrease of cloudiness – in the southern part of the region. The same positive significant correlation can be detected between geographic latitude and altocumulus, cumulonimbus and stratocumulus clouds, which indicates a decrease of these types of clouds in the southern part of the region, the figure being inverse in the northern parts. At the same time, stratus clouds have a negative significant correlation with latitude, decreasing in parallel with higher latitudes and showing an increasing trend in southern parts opening to the Black Sea (Fig. 6.3b).

The relation between geographic longitude and cloud change is not as obvious as in the case of geographic latitude and cloudiness; only cirrus clouds show a significant negative correlation with geographic longitude (Table 6.3), which means more cirrus clouds in the western part of the region and less cirrus clouds in the eastern part (Fig. 6.3c). The relation of cloudiness and orography is evident only in the case of cumulus clouds; they show a significant negative correlation, indicating a decrease in cumulus clouds with altitude in parallel with global warming (Fig. 6.3d).

6.5 Conclusions

In this study, regional changes in total cloudiness and a further eight types of clouds (Ac, As, Ci, Cb, Cu, Ns, Sc and St) in the south-eastern part of Europe are calculated in association with climate change. In the case of total cloudiness, the sign of change is mainly negative, meaning there is less cloudiness parallel to a 0.5°K increase in northern hemispheric temperature. Altostratus, cirrus, nimbostratus, stratocumulus and stratus clouds show the same behaviour as total cloudiness, i.e. a decrease with the warming climate. In contrast, altocumulus clouds show a high positive value, indicating an increasing trend with global warming. Cumulonimbus and cumulus follow the same trend as altocumulus, but with lower values.

Negative trends in regard to total cloudiness in the region fully coincide with related empirical and model estimations. In the neighbouring Carpathian basin, Mika (1988) found a significant increase in the duration of sunshine in the summer half of the year parallel to global warming, corresponding to a decrease in visual cloudiness, and no significant changes in the half year of winter. This estimation is based on a regression technique used on local series for over 70 years.

On the other hand, contemporary Global Circulation Models (GCMs) unequivocally indicate a few percent decrease in the annual mean total cloudiness between model years 2080–2099 vs. 1980–1999 (Meehl et al. 2007). Though GCMs are not capable of simulating individual cloud types, it is pointed out in the above-cited chapter of the IPCC report that less cloudiness is expected practically in the whole troposphere at given northern latitudes between the two time slices.

Differences in above-established trends between strati-form and cumuli-form of clouds can be interpreted using general physical considerations. A decrease in the first group might be connected to decreased relative humidity parallel to unequivocal warming. At the same time, convective clouds gain extra energy from the increasing vertical lapse rate yielded by maximum near-surface warming with a vertical profile cooling in the stratosphere (Trenberth et al. 2007).

The three analysed geographic factors, namely altitude, latitude and longitude partially explain the spatial distribution of changes in different cloud types. Total cloudiness, Ac, Cb and Sc show a significant positive correlation to geographic latitude within the study area. In the case of St, the correlation with geographic latitude is also significant but negative. Only the change in cirrus clouds exhibits a negative correlation with geographic longitude, and only changes in cumulus clouds depend statistically on altitude. Any attempt to explain these differences, however, would require regional climate modelling of the given area.

Acknowledgment The author hereby expresses most sincere thanks to Dr. János Mika at the Hungarian Meteorological Service for providing the idea of the study and method development.

References

Hahn CJ, Warren SG (1999) Extended edited synoptic cloud reports from ships and land stations over the globe 1952–1996. Internet publication and data http://cdiac.esd.ornl.gov/epubs/ndp/ndp026c/ndp026c.html

Jones PD (1994) Hemispheric surface air temperature variations: A reanalysis and an update to 1993. J Clim 7:1794–1802.

Loon Van H, Williams J (1976) The connection between trends of mean temperature and circulation at the surface, Part I, II and III. Mon Weather Rev 104:365–380.

Meehl GA, Stocker TF, Collins WD, Friedlingstein P, Gaye AT, Gregory JM, Kitoh A, Knutti R, Murphy JM, Noda A, et al. (2007) Global climate projections. In: Solomon SD, Qin D, Manning M, Chen Z, Marquis M, Averyt KB, Tignor M and Miller HL (eds) Climate Change 2007: The Physical Science Basis. Contribution of Working Group I to the Fourth Assessment Report of the Intergovernmental Panel on Climate Change, Cambridge University Press, Cambridge.

Mika J (1988) Regional features of global warming in Hungary. Idojárás 92:178–189, (in Hungarian).

Trenberth KE, Jones PD, Ambenje P, Bojariu R, Easterling D, Klein Tank A, Parker D, Rahimzadeh F, Renwick JA, Rusticucci M, et al. (2007) Observations: Surface and atmospheric climate change. In: Solomon SD, Qin D, Manning M, Chen Z, Marquis M, Averyt KB, Tignor M and

Miller HL (eds) Climate Change 2007: The Physical Science Basis. Contribution of Working Group I to the Fourth Assessment Report of the Intergovernmental Panel on Climate Change, Cambridge University Press, Cambridge.
Vinnikov KYa (1986) Sensitivity of climate. Gidrometeoizdat 219 p. (In Russian)
Ward JH (1963) Hierarchical grouping to optimizc an objective function. J Am Stat Ass 58:236.

Chapter 7
Research on 21st Century Climate Change in the Republic of Macedonia

Suzana Alcinova Monevska and Pece Ristevski

Abstract This chapter comprises updated results from research on projected climate change in the Republic of Macedonia in the course of the twenty-first century in relation to the previously set-out National Communications on Climate Change, an obligation under the United Nation Framework Convention on Climate Change. The results displayed here deal with air temperature and precipitation analysis on a seasonal and annual basis according to various emission scenarios for the years 2025, 2050, 2075 and 2100. Emission scenarios IS92a and IS92c as well as HadCM2, UKTR, UKHI-EQ, CSIRO-EQ and CCC-EQ models (for the First National Communication) and emission scenarios SRES A1T, A1Fl, A1B, and B1 (for the Second National Communication) are used. Projections are based on interpolated values from four selected Global Circulation Models (GCMs), namely CSIRO/Mk2, HadCM3, ECHAM4-OPYC3 and NCAR-PCM, whereas for local projection regarding different climatic regions of Macedonia, the method of empirical downscaling is applied.

The GCMs directly involving output on projected climate change in the Republic of Macedonia show a more intensive increase in air temperature in the summer season over the winter season and much higher values of expected temperature change in the country than the expected global temperature change. In general, almost no change in precipitation is expected for the winter season, but quite a strong decrease is expected in summer precipitation. Local projections for climate change indicate that different climatic regions in the country will respond differently to large-scale climate changes.

Keywords Climate change · Climate change scenarios · National communication

---

S.A. Monevska (✉)
Hydrometeorological Institute, Skopje, Republic of Macedonia
e-mail: smonevska@meteo.gov.mk

V. Alexandrov et al. (eds.), *Global Environmental Change: Challenges to Science and Society in Southeastern Europe*, DOI 10.1007/978-90-481-8695-2_7,

7.1 Introduction

The Republic of Macedonia ratified the United Nations Framework Convention on Climate Change (UNFCCC) on 4th December 1997. As a party to the convention, there is an obligation for regular preparation of national communications on climate change. The first national communication on climate change was prepared in 2003 in accordance with guidelines set by the Conference of the Parties for countries not included in Annex I of the convention. From March 2005 to March 2008, the Second National Communication on Climate Change was elaborated, in which the part dedicated to climate observations and research consists of the following:

- Basic characteristics and factors influencing climate in the Republic of Macedonia (Filipovski et al. 1996; Lazarevski 1993);
- Analysis, interpretation and comparison of long-term climatological data for the periods 1961–1990 and 1971–2000;
- Climate variations in the period 1926–2005 based on average air temperature data for measuring sites Skopje, Stip, Bitola, Prilep, Demir Kapija, Lazaropole and Ohrid;
- Climate variations in the period 1926–2005 based on annual precipitation values for measuring sites Skopje, Stip, Bitola, Prilep, Demir Kapija, Lazaropole and Ohrid.

Both national communications also comprise elaboration of different scenarios for changes in main climatic elements – air temperature and precipitation – in the twenty-first century.

7.2 Climate Change Scenarios for the Territory of the Republic of Macedonia in the First National Communication on Climate Change

Six climate models are used for investigation of climate change in the twenty-first century according to software package MAGICC SCENGEN and MAGICC (version 2.4 of 2000), published by the Intergovernmental Panel on Climate Change (IPCC), with which assessment of average values and values of low and high climate sensitivity are performed (Ministry of Environment and Physical Planning 2003). The climate models used in the analysis are: HadCM2 and UKTR of the UK' Hadley Center, UKHI-EQ of the Meteorological Service of the UK, CSIRO1-EQ and CISIRO2-EQ of the Australian Scientific Investigation Institute, and CCC-EQ of the Canadian Climate Center. On the basis of the above-mentioned IPCC scenarios, predictions of main climate elements (air temperature and precipitation) are made for the following years in the twenty-first century: 2025, 2050, 2075 and 2100.

Table 7.1 Changes in air temperature (°C) in the twenty-first century, according to emission scenario IS92a for four seasons (DJF-winter, MAM-spring, JJA-summer and SON-autumn)

	DJF				MAM				JJA				SON				ANNUAL			
	2025	2050	2075	2100	2025	2050	2075	2100	2025	2050	2075	2100	2025	2050	2075	2100	2025	2050	2075	2100
Low	0.7	1.2	1.7	2.2	0.6	1.0	1.4	1.8	0.8	1.4	1.9	2.4	0.7	1.3	1.8	2.3	0.7	1.2	1.7	2.2
Mean	1.0	1.8	2.5	3.3	0.8	1.4	2.4	2.6	1.1	2.0	2.8	3.6	1.2	2.0	2.9	3.7	1.0	1.7	2.5	3.2
High	1.4	2.5	3.6	4.7	1.1	2.0	2.9	3.8	1.5	2.8	4.0	5.2	1.4	2.5	3.7	4.8	1.4	2.5	3.5	4.6

Table 7.2 Changes in air temperature (°C) in the twenty-first century, according to emission scenario IS92c for four seasons (DJF-winter, MAM-spring, JJA-summer and SON-autumn)

	DJF				MAM				JJA				SON				ANNUAL			
	2025	2050	2075	2100	2025	2050	2075	2100	2025	2050	2075	2100	2025	2050	2075	2100	2025	2050	2075	2100
Low	0.6	0.9	1.1	1.2	0.5	0.7	0.9	0.9	0.7	1.0	1.2	1.3	0.6	0.9	1.1	1.2	0.6	0.9	1.1	1.1
Mean	0.9	1.3	1.6	1.7	0.7	1.1	1.3	1.4	0.9	1.4	1.8	1.9	0.9	1.3	1.6	1.8	0.8	1.3	1.6	1.7
High	1.2	1.9	2.3	2.6	1.0	1.5	1.9	2.1	1.3	2.1	2.6	2.9	1.2	1.9	2.4	2.7	1.2	1.8	2.3	2.5

Table 7.3 Changes in precipitation during the twenty-first century, according to emission scenario IS92a for four seasons (DJF-winter, MAM-spring, JJA-summer and SON-autumn)

	DJF				MAM				JJA				SON				ANNUAL			
	2025	2050	2075	2100	2025	2050	2075	2100	2025	2050	2075	2100	2025	2050	2075	2100	2025	2050	2075	2100
Low	1	1	2	2	–1	–1	–2	–2	–4	–7	–9	–12	–1	–2	–2	–3	–1	–2	–2	–3
Mean	1	2	2	3	–1	–2	–3	3	–5	–10	–14	–18	–1	–2	–3	–4	–1	–2	–3	–4
High	1	2	4	5	–1	–2	–4	–5	–8	–13	–19	–25	–2	–3	–5	–6	–2	–3	–5	–6

Table 7.4 Changes in precipitation during the twenty-first century according to emission scenario IS92c for four seasons (DJF-winter, MAM-spring, JJA-summer and SON-autumn)

	DJF				MAM				JJA				SON				ANNUAL			
	2025	2050	2075	2100	2025	2050	2075	2100	2025	2050	2075	2100	2025	2050	2075	2100	2025	2050	2075	2100
Low	1	1	1	1	–1	–1	–1	–1	–3	–5	–6	–6	–1	–1	–1	–2	–1	–1	–2	–2
Mean	1	1	2	2	–1	–1	–2	–2	–5	–7	–9	–9	–1	–2	–2	–2	–1	–2	–2	–2
High	1	2	2	3	–1	–2	–2	–3	–6	–10	–13	–14	–2	–2	–3	–3	–2	–3	–3	–4

Climate change scenarios for Macedonia are made supposing that the current policy of increasing the CO2 concentrations will be kept (Bergant 2006). Two projections of socioeconomic development (emission scenarios) have been chosen – IS92a and IS92c – whereby IS92a is "the best estimation" scenario of climate sensibility and IS92c is a scenario of "low" climate sensibility. Research results are given in the tables below. Significantly worse results are obtained according to the IS92a emission scenario: average annual increase of air temperatures is in the range of between 2.5°C in 2075 and 3.2°C in 2100. For the same scenario, the average sum of precipitation will decrease between –3% in 2075 and –4% in 2100 compared to the 1961–1990 period. With regard to emission scenario IS92c, the average annual air temperatures will increase in a range of 1.6°C in 2075 to 1.7°C in 2100, and the average sum of precipitation will decrease between –2.2% in 2075 and –2.4% in 2100 in comparison to the 1961–1990 period.

Tables 7.1 and 7.2 present the results of expected air temperature changes for two characteristic emission scenarios (IS92a and IS92c) as they show expected values for the twenty-first century in spring (March, April and May), summer (June, July and August), autumn (September, October and November) and winter (December, January and February) months respectively. Annual average air temperature increases will appear in the ranges of around 0.8°C (in spring), 1.1°C (in summer), 1.0°C (in autumn) and 1.0°C (in winter) for 2025; around 1.4°C (in spring), 1.9°C (in summer), 1.8°C (in autumn) for 1.8°C (in winter) for 2050; around 2.6°C (in spring), 3.5°C (in summer), 3.3°C (in autumn) and 3.2°C (in winter) for 2100. These results are obtained on the basis of emission scenario IS92a, whereas according to emission scenario IS92c, average air temperatures in 2100 will increase in the range of around 1.4°C (in spring), 1.9°C (in summer), 1.8°C (in autumn) and 1.7°C (in winter).

Precipitation in the twenty-first century for emission scenarios IS92a and IS92c will change as shown in Tables 7.3 and 7.4. According to the six above-mentioned models of climate change for the twenty-first century, decreases in precipitation will appear in the spring, summer and autumn, while in the winter precipitation quantities will increase. Decrease and increase amounts presented in Tables 7.3 and 7.4 are in respective ranges given in percentage of average sums during the period 1961–1990. It can be noted that the most significant changes will appear during summer months according to emission scenario IS92a.

7.3 Results of Climate Change Scenarios According to the Second National Communication on Climate Change

To prepare the Second National Communication on climate change, calculations were made to estimate main meteorological variables for the entire territory of the Republic of Macedonia for the years 2025, 2050, 2075 and 2100. The method of statistical downscaling has been applied for regional estimations of climate change. Long-term quality data records have been selected from different meteorological stations for climate and geographical regions (Ministry of Environment and Physical Planning 2008).

7.3.1 *Results of Direct GCM Output for the Entire Territory of the Country – First Estimate*

Projected changes in average daily air temperature (°C) and precipitation (%) for Macedonia are based on direct GCM outputs interpolated for the geographic location 21.5° E and 41.5° N covering the period 1961–1990. Values are presented separately for different seasons and are based on projections of results from four GCMs (CSIRO/Mk2, HadCM3, ECHAM4/OPYC3, NCAR-PCM) scaled to six emission scenarios (SRES A1T, A1Fl, A1B, A2, B1, and B2). The mean values are calculated across different emission scenarios and different GCMs; low/high values are respectively minimum/maximum across different scenarios and averaged across different GCMs.

As a first estimate of expected climate change over the territory of the entire country, the results of selected GCMs are interpolated for the geographic location 21.5° E and 41.5° N, i.e., approximately the middle of the country. The results of such an approach, called direct GCM output, show the highest increase in air temperature up to the end of this century to be in the summer season, together with the most intensive decrease in precipitation. In the case of precipitation, practically no change is expected in winter but a decrease is predicted in all other seasons. Details on direct GCM output projections can be found in Table 7.5.

A stronger air temperature increase in summer compared to other periods of the year could be related to the expected decrease in precipitation, because precipitation has a cooling effect on near-ground temperature in the summer months. The average temperature is likely to increase if precipitation decreases, and the temperature range is also likely to enlarge in the case of less precipitation and more sunny days. One has to also be aware that providing precipitation changes in percentages, which is common in climate change studies, has some caveats. Changes for the summer period might seem more dramatic in comparison to those for autumn and spring due to the fact that there is already little precipitation in most regions of Macedonia in the summer months. Consequently, the same absolute change presented in absolute values for summer and autumn can result in a much different relative value presented in %.

In the case of direct GCM output, additional projections for changes in scalar wind speed (in m/s) and incident solar radiation (Srad, in W/m^2) have been performed (not shown here). For both variables, relative expected changes are very small, practically not exceeding 10% in any direction, when considering the expected range. A minor increase in Srad is expected in all seasons, with the highest being in the summer. The generally small increase in incident solar radiation in all seasons corresponds to projected precipitation changes also showing the highest decrease to be in summer. Less precipitation means more clear days and thus more solar radiation reaching the ground. Practically no change is expected in wind speed over Macedonia according to direct GCM output for the four GCMs.

The direct GCM output method only provides a rough description of expected climate change in the twenty-first century in Macedonia and mostly deals with changes in average conditions over the entire country. The spatial variability of

Table 7.5 Projected changes in average daily air temperature (°C) and precipitation (%) for the Republic of Macedonia

	DJF				MAM				JJA				SON				ANNUAL			
	2025	2050	2075	2100	2025	2050	2075	2100	2025	2050	2075	2100	2025	2050	2075	2100	2025	2050	2075	2100
Average temperature (°C)																				
Low	0.7	1.4	1.8	2.2	0.7	1.3	1.8	2.2	1.2	2.2	3.2	3.7	0.8	1.5	2.2	2.6	0.9	1.6	2.2	2.7
Mean	0.8	1.7	2.3	3.0	0.8	1.5	2.2	3.2	1.4	2.5	4.1	5.4	0.9	1.7	2.8	3.7	1.0	1.9	2.9	3.8
High	0.9	1.9	2.9	4.2	0.9	1.8	2.9	4.6	1.7	2.9	5.1	7.6	1.1	2.0	3.6	5.3	1.1	2.1	3.6	5.4
Precipitation (%)																				
Low	1	5	3	4	–3	–2	–7	–5	2	–16	–21	–21	2	–2	0	–5	–1	–2	–4	–5
Mean	0	1	2	–1	–5	–6	–10	–13	–7	–17	–27	–37	–1	–4	–9	–13	–3	–5	–8	–13
High	–2	–1	1	–3	–7	–10	–13	–22	–24	–18	–33	–53	–3	–7	–17	–23	–6	–7	–12	–21

meteteorological parameters, as well as heterogeneous climate regions and country topography should also be kept in mind. Thus, the direct GCM output approach can present a benchmark for more detailed and complicated methods like empirical downscaling for the purpose of regional projection of future climate change.

7.3.2 *Results of the Empirical Downscaling Method*

Generally, the results of empirical downscaling show a greater increase in air temperature in winter and summer and less decrease in precipitation in winter than results estimated with GCMs. Projections for temperature and precipitation are compatible for the other seasons. The obtained values are based on results from projections of four GCMs (CSIRO/Mk2, HadCM3, ECHAM4/OPYC3, NCAR-PCM) scaled to six emission scenarios (SRES A1T, A1Fl, A1B, A2, B1, and B2) and presented separately for different seasons with regards to the reference period 1961–1990. Again, mean values are calculated as averages across different emission scenarios and different GCMs.

7.3.2.1 Southeastern and Central Parts of the Republic of Macedonia/sub-Mediterranean Climate Regions

When empirical downscaling projections for the region of southeastern Macedonia – with prevailing sub-Mediterranean climate represented by stations Gevgelija and Nov Dojran, and for central parts of Macedonia – with combinated continental and sub-Mediterranean climate, represented by stations Veles, Skopje–Petrovec, Strumica and Štip, are compared, a less intensive temperature change is evident for the first region in the winter, becoming more intensive in summer and autumn. Changes in air temperature in the spring are comparable in both sub-regions and the highest increase in air temperature is expected in summer. The differences between winter and summer increases in air temperature are especially evident for the southeastern region. Expected changes in precipitation are similar for both sub-regions as practically no change in precipitation is expected in the winter season and a decrease in precipitation looks likely in all other seasons (Table 7.6).

7.3.2.2 Southern and Southwestern Parts of the Republic of Macedonia/Continental Climate Region

Both the southern (represented by stations Bitola and Prilep) and southwestern (represented by stations Ohrid and Resen) parts of Macedonia are under prevailing continental climate influences. Climate change projections for these two regions are quite different, although the areas themselves are not far from each other. In the case of the southern region – represented by Bitola and Prilep – precipitation projections are very similar to regions with prevailing or partial sub-Mediterranean climate influence.

Table 7.6 Projected changes in average daily air temperature (°C) and precipitation (%) for central parts of Macedonia under a combination of sub-Mediterranean and continental climate impacts (represented by locations Veles, Strumica, Skopje-Petrovec, Štip) and for the southeastern part of Macedonia under sub-Mediterranean climate impacts (represented by locations Gevgelija and Nov Dojran)

	DJF				MAM				JJA				SON				ANNUAL			
	2025	2050	2075	2100	2025	2050	2075	2100	2025	2050	2075	2100	2025	2050	2075	2100	2025	2050	2075	2100
T	1.0	2.3	3.2	4.3	1.1	2.1	3.1	4.3	1.4	2.6	4	5.4	1.0	1.8	3.0	3.9	1.1	2.2	3.3	4.5
RR	0	1	2	–1	–5	–7	–11	–14	–6	–11	–18	–23	–1	–6	–11	–17	–3	–6	–9	–13
T	1.0	2.1	2.9	3.8	1.1	2.1	3.1	4.3	1.5	2.9	4.5	6.0	1.1	2.0	3.3	4.3	1.2	2.3	3.4	4.6
RR	–2	0	–1	–3	–6	–8	–13	–17	–4	–9	–14	–19	1	–4	–9	–14	–3	–5	–9	–12

Table 7.7 Projected changes in average daily air temperature (°C) and precipitation (%) for the southern part of Macedonia under continental climate impacts (represented by locations Bitola and Prilep) and for the southwestern part of Macedonia, also under continental climate impacts (represented by locations Ohrid and Resen)

	DJF				MAM				JJA				SON				ANNUAL			
	2025	2050	2075	2100	2025	2050	2075	2100	2025	2050	2075	2100	2025	2050	2075	2100	2025	2050	2075	2100
T	1.2	2.7	3.9	5.3	1.2	2.3	3.4	4.8	1.5	2.7	4.3	5.7	1.1	2.1	3.4	4.5	1.2	2.5	3.8	5.1
RR	–1	–1	–1	–3	–5	–7	–11	–14	–5	–12	–17	–22	–1	–5	–10	–15	–3	–5	–9	–13
T	0.9	2.0	2.9	3.9	1.0	1.9	2.9	4.1	1.1	2.0	3.1	4.2	0.9	1.6	2.7	3.5	0.9	1.9	2.9	3.9
RR	2	3	7	5	–5	–5	–9	–12	–3	–9	–13	–18	–2	–5	–10	–15	–2	–3	–5	–8

Almost no change in precipitation is expected in winter and and a slight decrease is foreseen for the other seasons, the strongest being in summer. Slightly stronger temperature changes are expected for this region compared to regions with a sub-Mediterranean climactic influence. Differences are especially evident in projections for the winter period. Contrarily, projections of temperature changes for the south-western region – represented by Ohrid and Resen – are much lower than those for the region represented by Bitola and Prilep. A slight increase in precipitation is even expected for winter with an evident decrease in other seasons. The different response of these two regions to large-scale climate variability could be related to the proximity of large water bodies (lake Prespa and lake Ohrid) for the Resen and Ohrid stations (Table 7.7).

7.3.2.3 Eastern Part of Republic of Macedonia/Continental Climate Region and Northwestern Part of Macedonia/Alpine Climate Region

Berovo and Kriva Palanka stations are representative of the eastern part of Macedonia, with a prevailing continental climate influence. The annual pattern of expected temperature change in this region is similar to the pattern for the continental region in southern Macedonia, but the intensity of change is slightly lower. A comparison with Bitola and Prilep stations shows a slight increase in precipitation expected in the winter but a decrease in all other seasons, relatively most intense for the summer. An increase in the daily air temperature range is also expected for summer and autumn.

For all three climate sub-types under mountainous influence (mountain/continental, sub-alpine, alpine) found in the northwestern part of Macedonia and represented by the Lazaropole, Popova Šapka and Solunska Glava stations, projections for changes in air temperature and precipitation are very similar. An increase of a few percent in precipitation by the end of the twenty-first century is expected in winter and a more intense decrease in all other seasons. The expected air temperature change is the strongest in this region of the country. The highest increase in air temperature is expected here in summer but the difference between seasons is not large (Table 7.8).

7.4 Comparison of Direct GCM Output Results

Estimates for temperature and precipitation changes in the twenty-first century are more dramatic than estimates based on the IS92a and IS92c emission scenarios. The direction of expected changes (e.g., strongest increase in air temperature and greatest precipitation decrease both occurring in summer) is the same, but their intensity is different. The difference is probably related to the fact that IS92 emission scenarios proposed by the IPCC in 1995 are more optimistic than SRES scenarios proposed in 2001. This can also be seen in projections for global temperature change

Table 7.8 Projected changes in average daily air temperature (°C) and precipitation (%) for the eastern part of Macedonia under continental climate impacts (represented by locations Kriva Palanka and Berovo) and for the northwestern part of Macedonia under prevailing alpine impacts (represented by locations Lazaropole, popova Sapka, and Solunska Glava)

	DJF				MAM				JJA				SON				ANNUAL			
	2025	2050	2075	2100	2025	2050	2075	2100	2025	2050	2075	2100	2025	2050	2075	2100	2025	2050	2075	2100
T	1.1	2.4	3.4	4.6	1.1	2.1	3.1	4.3	1.3	2.5	3.9	5.2	1.0	1.9	3.2	4.2	1.1	2.2	3.4	4.6
RR	2	4	8	6	–4	–5	–9	–11	–4	–10	–14	–20	–2	–7	–11	–15	–2	–5	–7	–10
T	1.2	2.7	3.8	5.2	1.3	2.5	3.7	5.3	1.5	2.8	4.5	5.9	1.1	2.2	3.6	4.7	1.3	2.6	3.9	5.3
RR	2	4	7	5	–4	–4	–7	–10	–4	–9	–13	–18	–2	–5	–9	–14	–2	–3	–5	–8

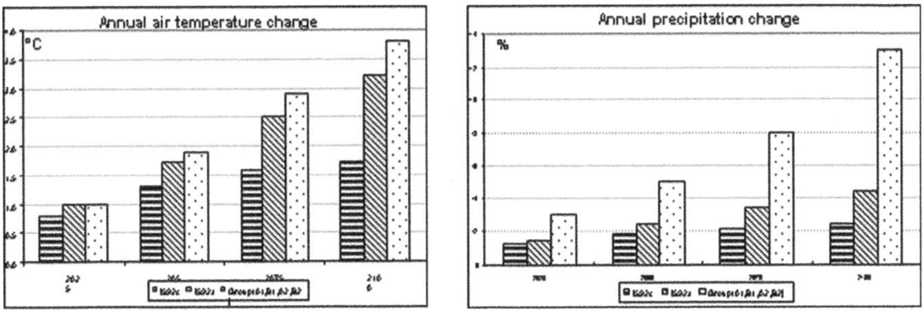

Fig. 7.1 Results comparison for different GCMs and scenarios

based on IS92 emission scenarios, which are lower than those based on SRES scenarios. Another reason could be different GCMs used in both studies (Fig. 7.1).

7.5 Conclusion

Evidently, valuable research has been carried out in the Republic of Macedonia under the UNFCCC in order to provide estimations for future changes in climate conditions for the twenty-first century. Significant progress has been achieved through the Second National Communication, in which not only general assessments of climate change for the territory of the entire country have been made in accordance with the latest GCMs and emission scenarios but also, first estimations for regional changes of basic climate parameters have been made.

The direct GCM output projected for Macedonia shows more intensive increases in air temperature in summer than in winter. Expected air temperature changes in the country during the twenty-first century are much higher than expected global temperature changes, but the results of our study are consistent with studies available for regions that include Macedonia. In general, almost no change in precipitation is expected for the winter season on the territory of Macedonia but quite a strong decrease is foreseen in the summer precipitation.

Local projections for climate change indicate that different climatic regions of Macedonia will respond slightly differently to large-scale climate changes. The continental climate region in southwestern Macedonia – close to the Ohrid and Prespa lakes – seems to have the weakest response to large-scale climate change in the sense of absolute temperature and precipitation changes, whereas the northwestern part – being under prevailing mountain/alpine climactic influences – shows the strongest response.

The differences obtained in GCM results show the need for further investigation and application of various methods and tools (e.g., PRECIS, dynamic downscaling, etc.) and for a critical review of present results about future climate change for the Republic of Macedonia. Although projections of local climate change are a step forward to achieving necessary knowledge about how different sub-regions of the Republic of Macedonia might respond to large-scale climate change, the many remaining uncertainties in respect to obtained data must be considered before such results are used in impact studies for human health, agriculture, forestry, biodiversity, water resources, energy management, etc. (Houghton et al. 2001; IPCC 1998, 2000)

References

Bergant K (2006) Climate Change Scenarios for Macedonia, Review of Methodology and Results, University of Nova Gorica, Slovenia.

Filipovski G et al. (1996) The Characteristic of the Climate–Vegetation–Soil Zones (Regions) in the Republic of Macedonia, Macedonian Academy of Sciences and Arts, Skopje.

Houghton JT et al. (2001) Climate Change. The Scientific Basis, Cambridge University Press, Cambridge.

IPCC (1998) A Special Report of IPCC Working Group II, The Regional Impacts of Climate Change, Cambridge University Press, Cambridge.

IPCC (2000) A Special Report of IPCC Working Group III, IPCC Special Report on Emissions Scenarios, Cambridge University Press, Cambridge.

Lazarevski A (1993) Climate of Macedonia, Kultura, Skopje.

Ministry of Environment and Physical Planning (2003) Macedonia's first national communication under the United Nations Framework Convention on Climate Change, Skopje.

Ministry of Environment and Physical Planning (2008) Macedonia's second national communication under the United Nations Framework Convention on Climate Change, Skopje.

Chapter 8
The Impact of Climate Change on the Precipitation Regime in Bosnia and Herzegovina

Dženan Zulum and Željko Majstorović

Abstract For the study presented below, observations from four stations in B&H (Sarajevo, Mostar, Tuzla and Bjelašnica) were considered. In order to examine differences from the past, and thus the progress of climate change, and since the station in Sarajevo has had unremitting terms from 1901 until now, data for the periods 1901–2006 and 1951–2006 are used for this station whereas for the other stations, only data from the period 1951–2006 are used. Maximum daily precipitation, i.e. one daily, two daily, three daily and five daily maximum precipitation incidents are studied. As well, maximum 10- and 5-year values and short-term rain for the above periods are examined. It is interesting to join the linear trends in these periods and thus ascertain increases in values, while at the same time there has been an increase in periods of drought. The situation is only different at the Bjelašnica station, where there has been a decrease in these values. Such results are especially startling for the year 2007, when through April-September, more extreme warm and extreme cold periods were observed, followed by drought and extreme precipitation events.

Keywords Linear trend · Climate change · Drought · Precipitation regime · Temperature

8.1 Introduction

Dominated by mountainous and hilly terrain, and drained by major rivers to the north (Sava) and east (Drina), Bosnia & Herzegovina has a climate that is as variable as the rest of the former Yugoslav federation, with moderate continental climatic conditions generally the norm. The climate features hot summers and cold winters. In higher elevations of the country, summers tend to be short and cold while winters tend to be long and severe. Along the coast,

D. Zulum (✉)
Federal Hydrometeorological Institute, Sarajevo, Bosnia and Herzegovina
e-mail: dzenanz@fhmzbih.gov.ba

V. Alexandrov et al. (eds.), *Global Environmental Change: Challenges to Science and Society in Southeastern Europe*, DOI 10.1007/978-90-481-8695-2_8,

winters tend to be short and rainy. In July, the mean temperature is 22.5°C. January's mean temperature is 0°C. Annual rainfall averages roughly 62.5 cm. The southern area of Herzegovina benefits from a Mediterranean climate, with an annual precipitation of 61–80 cm while the Central and Northern areas of Bosnia and Herzegovina has an Alpine climate with precipitation ranging from 150 to 249 cm.

In previous studies (e.g. Majstorović et al. 2004, 2005, 2008) some effects of climate change in Bosnia and Herzegovina are discussed, namely: – an increase in average temperature of about 0.7°C over the past 100 years; – no essential change (only ± 5% in the past 100 years depending on the region in the country) in the annual sum of rainfall. However, the existence of very warm and very cold short periods, and their consecutive fast exchanges have been documented, and this phenomenon has a negative effect on human health. The above-mentioned effects also impacts changes in the precipitation regime.

8.2 Data and Method

In the last decades the territory of Bosnia and Herzegovina was observed significant weather variability followed by the very warm and very cold short periods. It impacts on the regime of temperatures and precipitation. The decade 1997–2006 has been the warmest in the last 50 years for Bosnia and Herzegovina (B&H), and 2007 is the 5th warmest year since meteorological observations in the country were initiated (Hodžić and Voljevica 2008). These facts are important not only because of the high temperatures themselves, but because they indicate the existence of very warm and very cold short periods. Similar oscillations can be detected in terms of precipitation. To study these phenomena, data from the weather stations presented in Table 8.1 are used:

Table 8.1 Weather stations used in the study

Weather station	Latitude (φ)	Longitude (λ)	Altitude (m)
Mostar	43° 21′	17° 48′	99
Sanski Most	44° 46	16° 42′	158
Sarajevo	43° 52	18° 26′	630
Tuzla	44° 33	18° 42′	305
Bjelašnica	43° 43	18° 16′	2,067

The impact of climate changes in the precipitation regime of Bosnia and Herzegovina is studied through linear trends in corresponding graphs (Figs. 8.1 and 8.2). Maximum precipitation for 1, 2, and 5 days in the period 1951–2006, with an overview of the year 2007, is examined. The index of De Martonne was calculated, as well.

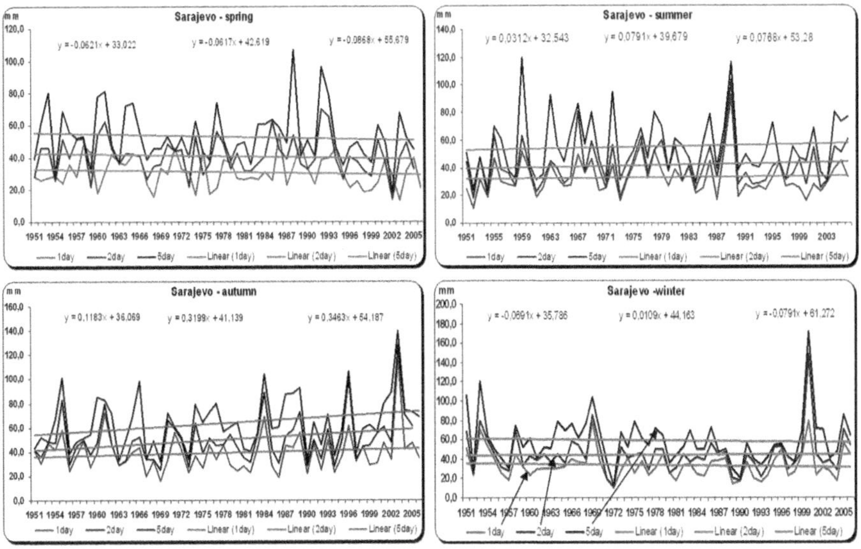

Fig. 8.1 Maximum precipitation in Sarajevo for 1, 2, and 5 days in the period 1951–2006, during different seasons

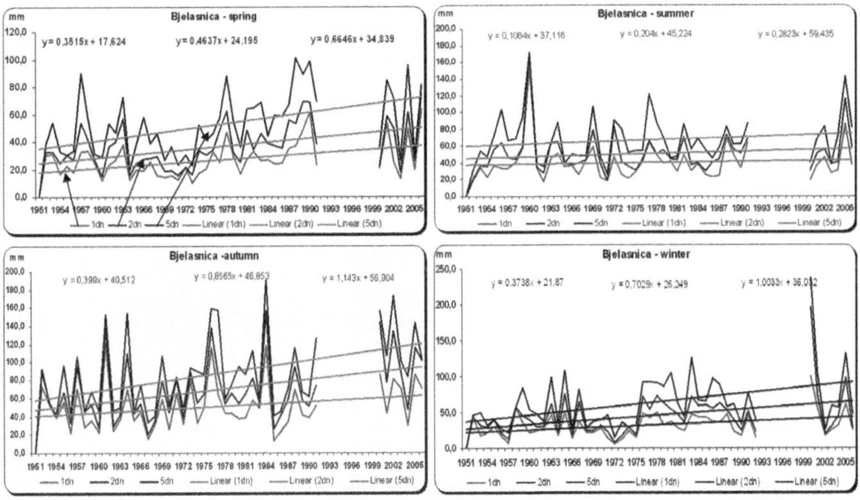

Fig. 8.2 Maximum precipitation in Bjelašnica for 1, 2, and 5 days in the period 1951–2006, during different seasons

8.3 Results

Annual and seasonal results are presented in Tables 8.2 and 8.3. Analyses of the precipitation regime indicates large increasing trends in precipitation intensity for most of Bosnia and Herzegovina, except for the region around station Sanski Most. Increases in precipitation intensity are observed for the region around stations Bjelasnica and Mostar (Mediterranean climate area), which could relate to an increased number of cases of Mediterranean air masses passing over the Dinarydes mountain. Table 8.3 shows an increasing trend in precipitation intensity for most parts of the country in the summer and autumn.

Table 8.3 shows also a decreasing trend in precipitation intensity for most parts of the country in the winter period, except in the region of Bjelasnica station. Large variability in precipitation regimes is seen. These findings are in harmony with the present world processes of global warming.

Drought periods are studied using the De Martonne Index (Table 8.4) for the period 1997–2006 and 2007 with average values of the same parameter from the

Table 8.2 Annual linear trend of maximum precipitation for 1, 2, and 5 days in the period 1951–2006

	Sarajevo	Bjelašnica	Mostar	Sanski Most	Tuzla
1 day	0.004575	0.307675	0.007925	−0.11613	0.060775
2 days	0.09645	0.556775	0.11745	−0.14405	0.1031
5 days	0.072925	0.7733	0.062325	−0.1665	0.127725

Table 8.3 Seasonal linear trend of maximum precipitation for 1, 2, and 5 days in the period 1951–2006

	Sarajevo	Bjelašnica	Mostar	Sanski Most	Tuzla
Spring					
1 day	−0.0621	0.3515	0.2607	−0.153	−0.0831
2 days	−0.0617	0.4637	0.2847	−0.1676	−0.0673
5 days	−0.0868	0.6646	0.1858	−0.2671	−0.155
Summer					
1 day	0.0312	0.1064	−0.0031	−0.1853	0.1897
2 days	0.0791	0.204	0.1757	−0.2753	0.2383
5 days	0.0768	0.2823	0.0597	−0.4608	0.2449
Autumn					
1 day	0.1183	0.399	0.037	−0.135	0.1578
2 days	0.3199	0.8565	0.2248	−0.097	0.2879
5 days	0.3463	1.143	0.3893	0.0215	0.4256
Winter					
1 day	−0.0691	0.3738	−0.2629	0.0088	−0.0213
2 days	0.0109	0.7029	−0.2154	−0.0363	−0.0465
5 days	−0.0791	1.0033	−0.3855	0.0404	−0.0046

Table 8.4 De Martone index for different months from 1997 to 2007

Year	III	IV	V	VI	VII	VIII	IX
	Sarajevo						
1997	37	88	29	20	33	23	9
1998	41	49	21	41	15	24	55
1999	44	58	30	31	40	33	47
2000	46	27	33	9	20	6	32
2001	42	74	34	54	26	28	121
2002	30	71	40	19	30	38	101
2003	2	20	22	38	25	3	38
2004	40	67	51	35	56	16	32
2005	69	43	54	35	44	61	49
2006	114	45	29	35	47	75	19
2007	55	11	51	23	25	10	75
	Sanski Most						
1997	38	87	30	46	41	27	25
1998	38	57	56	31	18	17	86
1999	22	63	34	45	46	13	52
2000	50	29	32	11	40	1	36
2001	39	69	16	65	21	3	141
2002	19	100	80	24	13	52	90
2003	29	14	25	17	26	9	59
2004	67	94	38	48	34	14	46
2005	58	37	40	30	42	62	39
2006	80	78	42	42	9	73	19
2007	56	9	74	38	19	31	102
	Mostar						
1997	7	94	28	10	13	26	7
1998	9	66	47	22	7	21	135
1999	80	92	36	28	31	20	31
2000	38	33	22	5	24	5	26
2001	93	66	14	20	10	1	97
2002	26	65	39	15	23	61	82
2003	0	34	27	19	9	7	36
2004	112	71	70	29	10	35	22
2005	102	59	38	12	37	49	73
2006	83	46	31	31	7	46	53
2007	93	7	38	19	8	3	42

period 1961–1990. The results show unusually large numbers of drought periods, ranging inside these from moderate to extreme drought. The De Martonne Index is low in the summer season, especially for the southwestern area of Bosnia and Herzegovina.

8.4 Conclusions

An analysis of the precipitation regime indicates a large increasing trend in precipitation intensity for most of Bosnia and Herzegovina. This is especially characteristic in the summer and autumn seasons. Evidently, there is an increased intensity in precipitation (rain showers) causing increased runoff and soil moisture deficit.

Study results allow us to conclude that climatic change in the last decade (1997–2006) has led to increased water requirements for crop growth and greater soil water deficit, as well as greater demand for larger quantities of suitable irrigation water. Inside decade values, the presence of cold and warm short periods as well as large variability in precipitation regimes are found. These findings are in agreement with the current international process of global warming (Morell 2001, Climate changes, Personal communication; Salinger 2005).

References

Hodžić S, Voljevica S (2008) Weather conditions during the winter season 2007/2008.

Majstorovic Z, Toromanović A, Gabela L (2005) Changes in precipitation regime in Sarajevo (1894–2003), ICAM-MAP, Zadar 2005. pp. 667–668.

Majstorovic Z, Toromanović A, Halilović F (2004) Trends of climatic changes considering over years 1894–1993 and 1894–2003 for Sarajevo, BALWOIS Conference, Ohrid 2004.

Majstorović Z, Zulum D, Ahmović I, Hadžismajlović I (2008) Warm and cold waves and precipitation variability on territory of BiH in the last ten years comparative with row 1961–1990 BALWOIS Conference, Ohrid.

Salinger J (2005) Climate variability and change: Past, present and future – an overview, Climatic Change 70(1–2), 9–29.

Chapter 9
Climatological Analysis of the Synoptic Situations Causing Torrential Precipitation Events in Bulgaria over the Period 1961–2007

Lilia Bocheva, Ilian Gospodinov, Petio Simeonov, and Tania Marinova

Abstract Variations in heavy and extreme precipitation are interesting to study as these events cause considerable damage and loss of life worldwide each year. The upward tendency of damages caused by natural disasters supports the idea that extreme events associated with the effects of climate change, such as torrential precipitation, are occurring with greater frequency. The same tendency is being observed in Bulgaria over the last decade of the twentieth century. Namely, there have been negative trends in annual and seasonal precipitation totals, but an increase in the contribution of heavy rainfall events to total precipitation. This pattern has also been observed for other Mediterranean countries.

In this chapter, a comparative analysis of some very extreme precipitation events is carried out using all available data for torrential precipitation (considered to be rain events totaling $\geq$ 100 mm/24 h at one station) from the meteorological network of the National Institute of Meteorology and Hydrology (NIMH) for the period 1961–2007. An increase of about 30% in the mean annual number of days with torrential precipitation is found for the period 1991–2007 versus the period 1961–1990. NIMH's historical archive of synoptic maps and NCEP/NCAR reanalysis data files are used for analysis and classification of synoptic situations causing torrential precipitation over the country. Respective fields of air pressure and wind velocity are also considered.

Keywords Bulgaria · Synoptic analysis · Torrential precipitation

9.1 Introduction

Studies of monthly, seasonal and yearly precipitation at local and global scales show certain trends in time over many regions of the world (Dai et al. 1997; Easterling et al. 2000). The trends are globally positive through the last century although some

L. Bocheva (✉)
National Institute of Meteorology and Hydrology, Sofia, Bulgaria
e-mail: Lilia.Bocheva@meteo.bg

V. Alexandrov et al. (eds.), *Global Environmental Change: Challenges to Science and Society in Southeastern Europe*, DOI 10.1007/978-90-481-8695-2_9,

areas have been characterized by negative trends. In most areas of the world, trends in common rainfall have the same sign as the trends in amounts of 1-day heavy rain. In Bulgaria, decreases in total seasonal and yearly precipitation have taken place (Alexandrov et al. 2004; Sharov et al. 2000), however an increase in extreme daily rainfall has also been observed (Bocheva et al. 2006, 2009).

Extreme precipitation events in Bulgaria are rare and local phenomena usually registered at one or more neighboring stations but not on a regular basis. Precipitation amounts equal to or greater than 100 mm/24 h measured at a certain station are not always connected with similar amounts at neighboring stations because they are usually produced by small scale weather phenomena and local severe convective storms. For these reasons, the existing meteorological network is not always suitable for observation of extreme weather events. During the last 5 years, a series of hazardous events, most often associated with severe convective storms and heavy rainfall, have affected Bulgaria. They have caused local floods, significant property damage and loss of life and have thus had a considerable impact on the Bulgarian economy.

The aim of investigations presented here is to study monthly and seasonal distribution of torrential precipitation events over the territory of Bulgaria. Precipitation totals equal to or greater than 100 mm/24 h at one station – nearly two times above the average monthly precipitation amount for all Bulgarian meteorological stations – are considered in the analysis. Classification by season and region of synoptic situations displaying this type of precipitation are made.

9.2 Data

Analysis is carried out on the basis of data about torrential precipitation events from the meteorological database of the Bulgarian National Institute of Meteorology and Hydrology (NIMH) for the period 1961–2007. Records from as many as 542 synoptic, climatological and rain-gauge stations, in which regular observations were completed during this whole period or part of it, are considered (Fig. 9.1). The period was chosen because instruments and methods of measurement for daily precipitation amounts were the same in all stations: the measured daily precipitation total was taken at 7.30 a.m. local time with classic ground-level precipitation gauges. Data from automatic stations were not included in this study.

In order to better present variations in extreme precipitation events, monthly and annual distributions of days with torrential precipitation over the whole country for two periods are summarized first: 1961–1990 (basic period) and 1991–2007, then once again for each of the decades 1961–1970, 1971–1980, 1981–1990, 1991–2000 and 2001–2007. The number of meteorological stations has changed during the entire investigation period, growing from 333 stations in 1961 up to 542 in 1992 and then dropping to 432 today (Fig. 9.2). A summary of the mean annual and mean monthly number of days with extreme precipitation for the whole country for two periods, 1961–1990 and 1991–2007, is made. Brief statistical analysis is applied

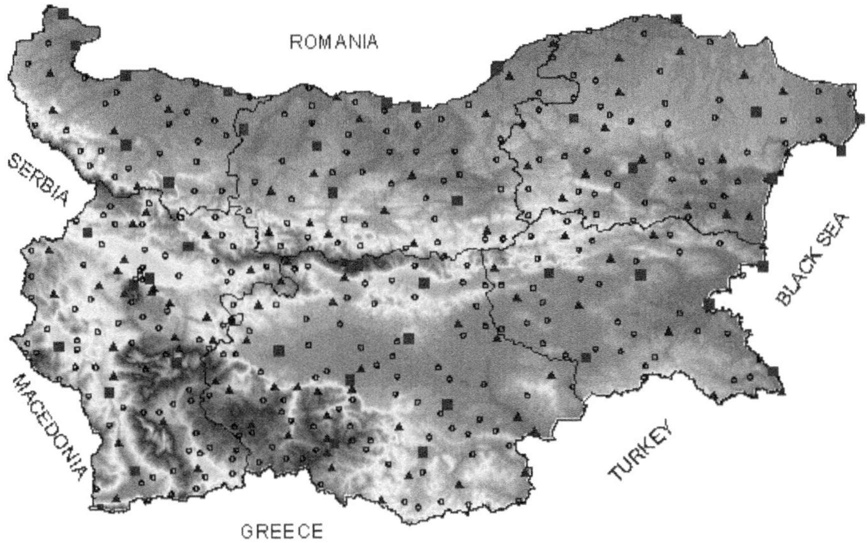

Fig. 9.1 NIMH meteorological network in Bulgaria: synoptic (*squares*), climatological (*triangles*) and rain-gauge (*circles*) stations

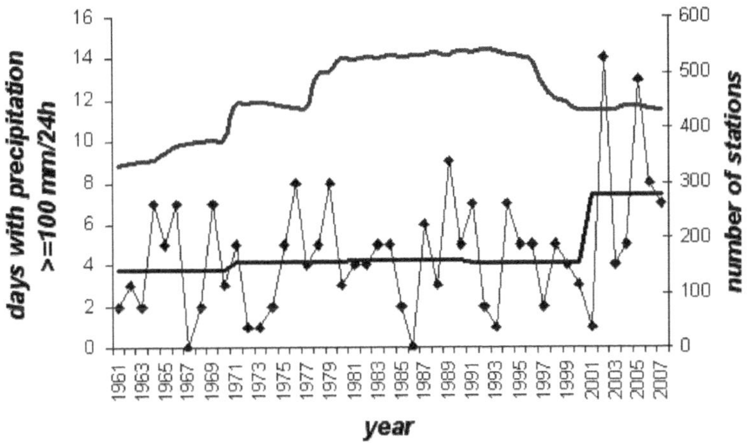

Fig. 9.2 Annual number of days with precipitation totals ≥ 100 mm/24 h, compared to the number of meteorological stations for the period 1961–2007

for the assessment of variability and possible differences in torrential precipitation from long-term data series. The Poisson distribution and corresponding nonparametric tests are applicable to such discrete samples of large-scale heavy rain days. Synoptic situations leading to such torrential 24-h precipitation are analysed using the NCEP/NCAR reanalysis data files and EUMETSAT images.

9.3 Torrential Precipitation in Bulgaria Between 1961 and 2007

During the 47-year period under study, annual distribution of the number of days with precipitation equal to or greater than 100 mm/24 h, registered in at least one station, showed a tendency to increase (Fig. 9.2). The mean annual number of days with hazardous precipitation during the period 1991–2007 is 5.5, with a standard deviation of 3.7, which in comparison with the 1961–1990 period – (4.2±2.4 days), presents an increase in mean annual extreme rainfall days of about 33% (Table 9.1). The comparison of mean annual numbers of days with torrential precipitation calculated for each decade (see Fig. 9.2 and Table 9.1) shows that the biggest contribution of heavy precipitation is due to the most recent period (2001–2007). While over the previous four decades the mean annual number of days has been almost the same (about 4 days), during the last 7-year period it has risen to about 7 days, representing an increase in the number of days with danger-inducing precipitation of about 80%.

Table 9.1 Statistical comparison between all samples of the average number of precipitation days (daily precipitation ≥100 mm at least in one station) using Poisson distribution for each previous decade (1) and one recent (2001–2007) (2) data set

No of sample	1 Mean μ_1	1 Min	1 Max	2 Mean μ_2	2 Min	2 Max	$1, 2\ \chi^2$	p teil probability	$(\mu_2 - \mu_1)/\mu_1$ (%)
1961/70 vrs 2001/07	3.8	0	7	7.4	1	14	10.03	0.0015	94.7
1971/80 vrs 2001/07	4.2	1	8	7.4	1	14	7.61	0.0058	76.2
1981/90 vrs 2001/07	4.3	0	9	7.4	1	14	7.07	0.0078	72.1
1990/00 vrs 2001/07	4.1	1	7	7.4	1	14	8.17	0.0043	80.5
1961/90 vrs 1991/07	4.1	0	9	5.5	1	14	4.33	0.0374	33.4

The inter-annual distribution of the mean number of days with extreme precipitation is presented in Fig. 9.3. A shift in the maximum distribution of heavy rain days for the two investigated periods can be found. During the period 1961–1990, the greatest number of heavy rain days took place in December and June. However, in more recent periods, such precipitation events occurred more frequently in July and September, when their increment rising by about 75–100%. On one side, there is an increase in the frequency of heavy rain episodes in all months from May to November in the period 1991–2007. On the other side, there is a recent tendency for decrease in extreme precipitation days during winter and especially during spring months (March and April). These results were also confirmed by Cavazos (2000) and Bocheva et al. (2009), supporting the downward trend of annual precipitation, and more particularly of winter precipitation over the Balkans, as obtained by Alexandrov et al. (2004), Sharov et al. (2000) and Sahsamanoglou et al. (1997).

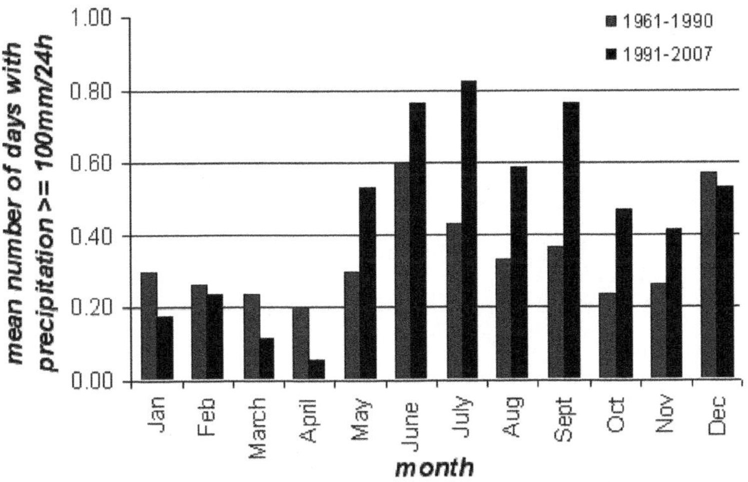

Fig. 9.3 Inter-annual distribution of the mean number of days with extreme precipitation

9.4 Weather Patterns Associated with Torrential Precipitation in BULGARIA

9.4.1 Autumn–Winter Patterns (October–March)

The areas most likely to receive torrential precipitation in these months are located in the West and East Rhodope Mountains.

9.4.1.1 West Rhodopes

In the region of the West Rhodopes, there is typically a large amount of rain or snow from the southwest mid-troposphere jet (SW-jet) traveling through Bulgaria. Alongside the jet, rapid Mediterranean cyclones pass through the country. They are either warm or cold depending on the exact geographic position of the waves alongside the thermal front associated with the SW-jet. The precipitation is mostly dynamic in nature due to the interaction of antagonistic cold and warm air masses additionally forced together by orography. The most typical structure is given in Fig. 9.4.

9.4.1.2 East Rhodopes

In the region of the East Rhodopes, it is typical to have large amounts of rain (and rarely snow) when a southern jet (S-jet) passes over Bulgaria. This is usually associated with a stationary or slowly moving deep Mediterranean cyclone to

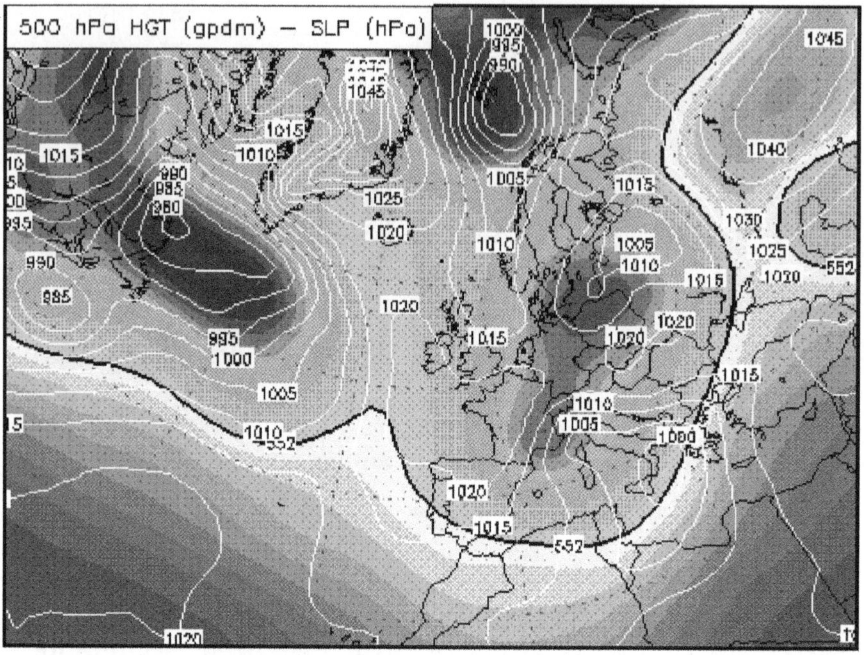

Fig. 9.4 H500 (*grey scale*) and SLP (*white contour*). 08.01.1985 12 h GMT

the southwest of Bulgaria, or with a cut-off cyclone in the same position. These are warm cyclones. The precipitation is mostly convective in nature due to forced convection through dynamic convergence together with orographic forcing and neighboring, relatively warm, sea waters. The most typical structure is given in Fig. 9.5. It is a very well pronounced pattern.

9.4.2 Late Winter–Early Spring (February–May)

At this time of year, the area most at risk for violent weather is the southwestern mountainous corner of Bulgaria. Such weather is mostly associated with a stationary or cut-off cyclone centered over Serbia-Albania-Greece which usually has a cold core. It can also be associated with an anti-cyclonic belt over Central Europe. The dominant mid-troposphere flow over Bulgaria comes from the southeast. This configuration brings precipitation of a mixed nature – both convective and dynamic. However, considerable amounts of rain or snow are definitely due to convection in late winter and early spring over mountainous triggers. The most typical structure is given in Fig. 9.6. It shows a less pronounced pattern.

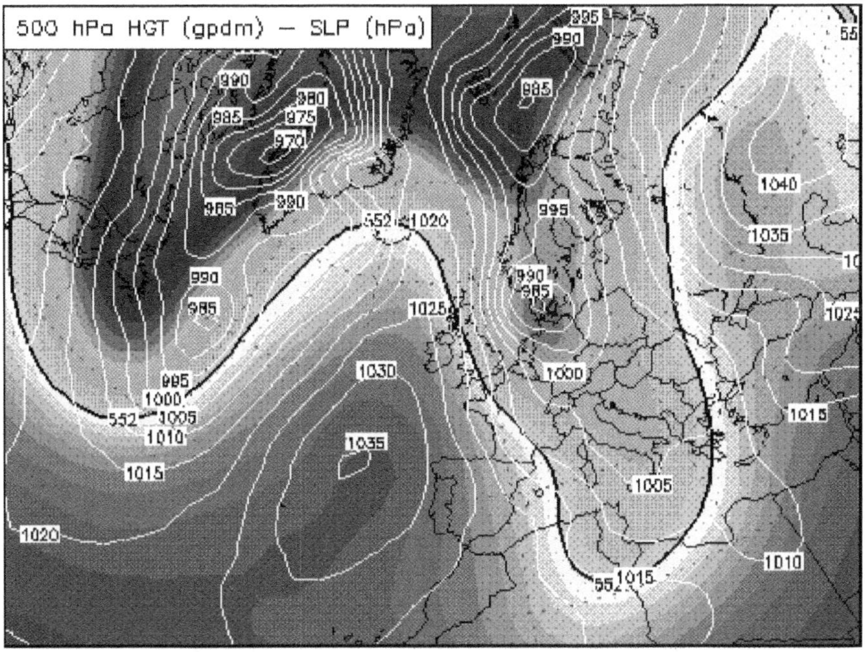

Fig. 9.5 H500 (*grey scale*) and SLP (*white contour*). 12.12.1990 12 h GMT

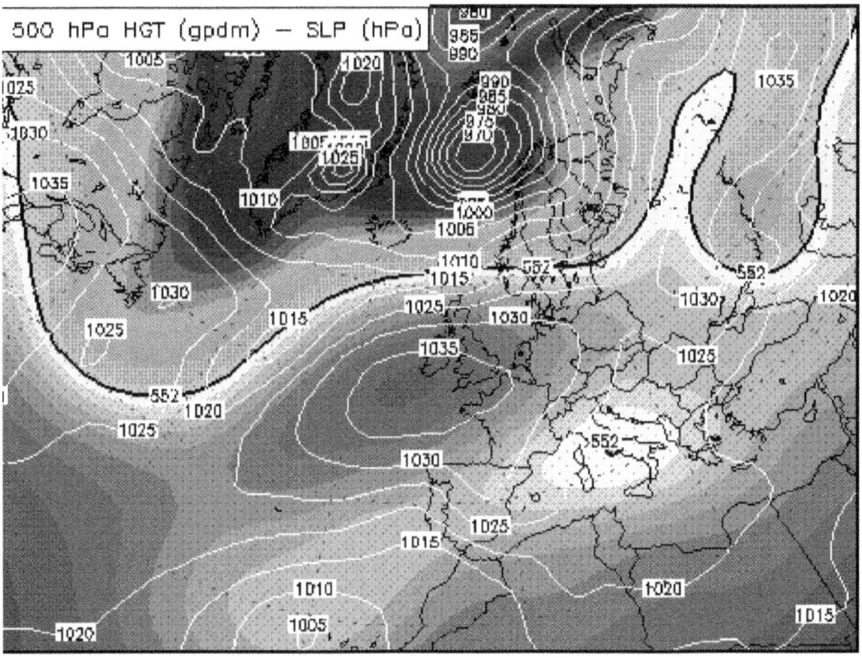

Fig. 9.6 H500 (*grey scale*) and SLP (*white contour*). 30.03.1990 00 h GMT

9.4.3 *Late Spring–Early Summer (May–July)*

In these months, the most at-risk area for torrential rain is the northwestern part of the country.

9.4.3.1 Central West and Northwestern Bulgaria

In these areas, convective precipitation is generally associated with a southwestern mid-troposphere flow over Bulgaria. Alongside such flow, meso-scale convective systems are generated over mountainous ridges, which then migrate northeastwards. Usually, there is a cyclonic area over Central Europe. The most typical structure is given in Fig. 9.7.

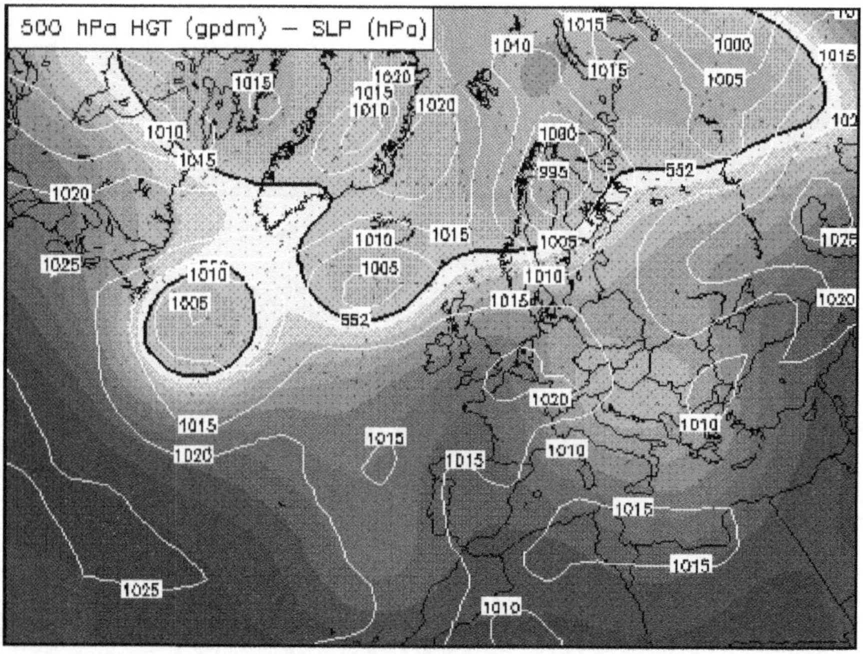

Fig. 9.7 H500 (*grey scale*) and SLP (*white contour*). 06.06.2004 00 h GMT

9.4.3.2 Central Northern Bulgaria

In this region, a weather pattern favoring heavy rain usually comes from a complex synoptic-scale structure. The result is favorable conditions for convergence in the central part of northern Bulgaria. The Stara Planina chain serves as a convection trigger within a mid-troposphere southern flow. The most typical structure is given in Fig. 9.8.

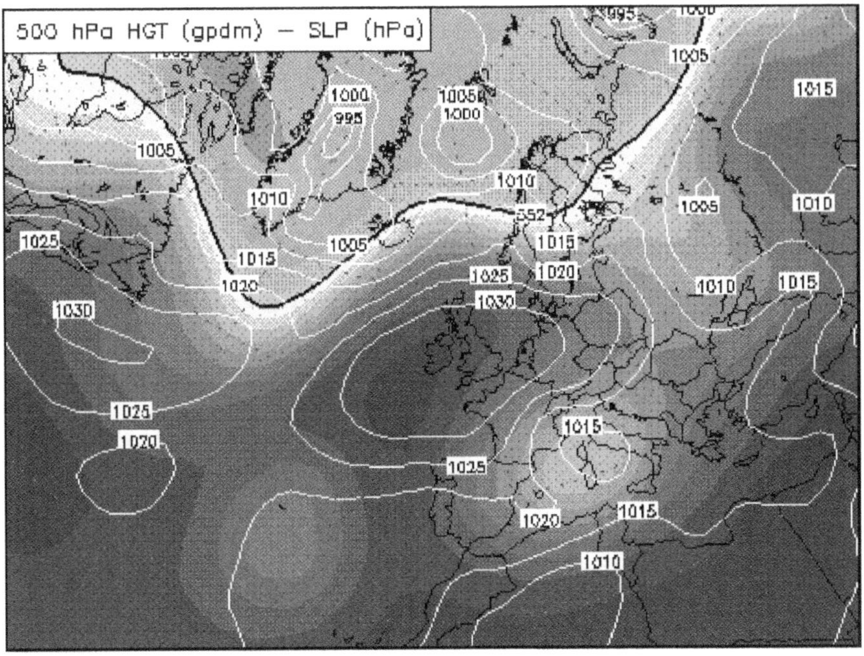

Fig. 9.8 H500 (*grey scale*) and SLP (*white contour*). 12.06.1994 12 h GMT

9.4.4 Late Summer–Autumn (July-November)

9.4.4.1 East Bulgaria and the Black Sea Coast

During these months, the area most in danger of torrential rains is eastern Bulgaria, especially the Black Sea coast. The southeast corner of the country is at particularly high risk, nearly always because of Black Sea cyclones. They are either stationary cut-off cyclones associated with a high-pressure belt over Central and Eastern Europe or part of a series of Mediterranean cyclones generated at the bottom of a deep trough over Eastern Europe. The large amounts of precipitation are due to the relatively warm waters of the Black Sea. The Strandzha region is particularly endangered if the cyclone is centered over Asia Minor. The most typical structure is given in Fig. 9.9.

9.4.4.2 The Mountainous Part of Southwestern Bulgaria and Other Central Parts of the Country

A cyclonic structure of intermediate size between a full synoptic scale cyclone and a meso-scale convective cyclonic system moves from the west to the east through the country. It is usually a cut from the polar front and is self-sustained through day-time convection. The neighboring warm sea water provides moisture. The most typical structure is given in Fig. 9.10a, together with corresponding infrared (IR)

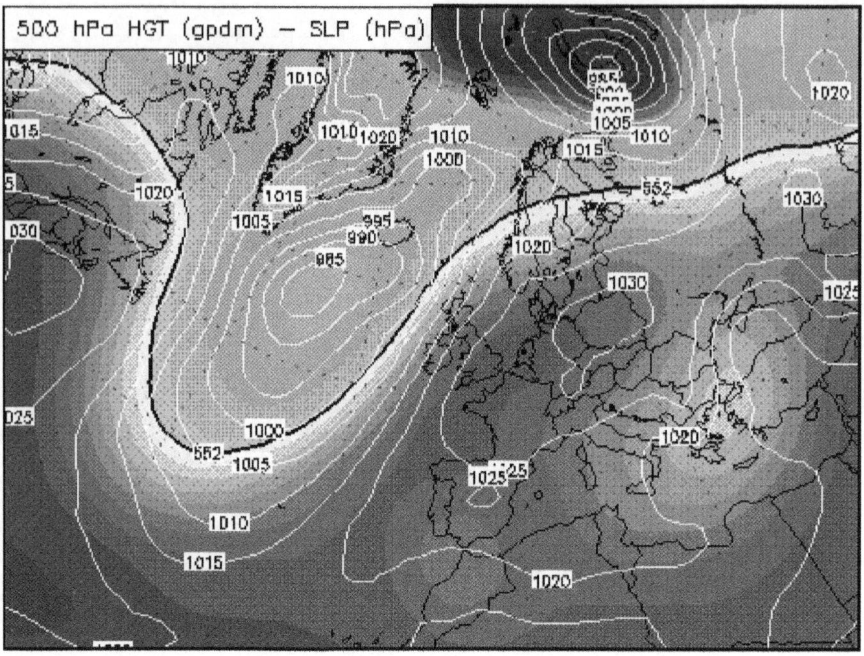

Fig. 9.9 H500 (*grey scale*) and SLP (*white contour*). 25.10.1975 00 h GMT

satellite image – Fig. 9.10b. It is a well-pronounced pattern, but the endangered area varies significantly from case to case.

9.5 Conclusions

The analysis of torrential precipitation events registered in the meteorological network of NIMH for the period 1961–2007 leads us to the conclusion that the annual distribution of days with precipitation ≥100 mm/24 h, observed at least in one station, reveals an upward trend. Similar results have been obtained by Nastos and Zerefos (2008), i.e. positive trends in extreme daily precipitation (totals exceeding 30 mm and 50 mm) in the eastern and southeastern regions of Greece during the period 1957–2001.

In Bulgaria, the mean annual number of days with torrential precipitation during the period 1961–1990 is 4.2 (±2.4) days – about 33% less in comparison with the period 1991–2007. The greatest number of heavy rain days for the period 1961–1990 is observed in December and June. In recent decades (1991–2007), this kind of precipitation more frequently occurs in July and September. The increase in the number of heavy rain days is about 75–100%. An increase in the frequency of heavy rain events from May to November is noted over the period 1991–2007, too.

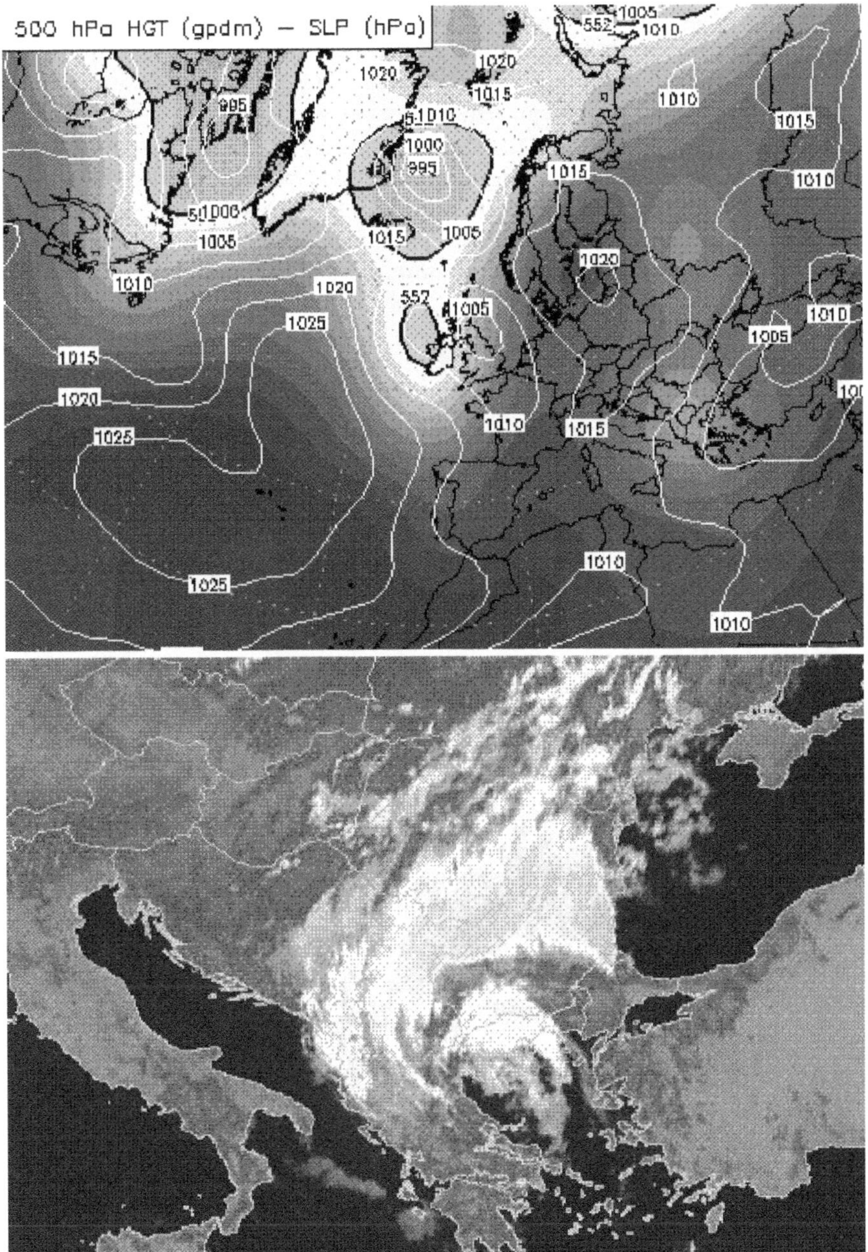

Fig. 9.10 (**a**) H500 (*grey scale*) and SLP (*white contour*) 06.08.2007 00 h GMT; (**b**) Meteosat IR satellite image 06.08.2007 12 h GMT

Contrarily, recent tendencies showing a decrease in extreme precipitation days have been observed during winter and especially spring months (March and April).

Extreme precipitation events over Bulgaria are clustered in space (by region) and time (by season) and the recognized clusters are associated with weather patterns. This first attempt at synoptic classification can be the basis of further investigations on the influence of climate changes on frequency and intensity increases in torrential precipitation events in different regions of Bulgaria. The obtained results can be of help in the development of methods for middle and long-range forecasting of such events as well as for assessment of the most probable damages.

References

Alexandrov V, Schneuder M, Koleva E, Moisselin J-M (2004) Climate variability and change in Bulgaria during the 20th century. Theor Appl Climatol 79:133–149.

Bocheva L, Simeonov P, Marinova T (2006) On frequency distribution and intensity of severe convective storms over Bulgaria. In: Proceedings of BALWOIS 2006, Republic of Macedonia, Ohrid (ISBN 9989-9594-1-2, CD version).

Bocheva L, Marinova T, Simeonov P, Gospodinov I (2009) Variability and trends of extreme precipitation events over Bulgaria (1961–2005. Atmos Res 93:490–497.

Cavazos T (2000) Using self-organizing maps to investigate extreme climate events: an application to wintertime precipitation in Balkans. J Clim 13:1718–1732.

Dai A, Fung IY, Del Genio AD (1997) Surface observed global land precipitation variations during 1900–1988. J Clim 10:2943–2962.

Easterling D, Meehl G, Parmesan C, Changnon S, Karl T, Mearns L (2000) Climate extremes: observations, modeling, and impacts. Science 289:2068–2074.

Nastos PT, Zerefos CS (2008) Decadal changes in extreme daily precipitation in Greece. Adv Geosci 16:55–62.

Sahsamanoglou H, Makrogiannis T, Hatzianastasiou N, Rammos N (1997) Long term change of precipitation over the Balkan Peninsula. In: Ghazi A et al. (eds) Eastern Europe and Climate Change, European Commission, Brussels.

Sharov V, Koleva E, Alexandrov V (2000) Climate variability and change. In: Staneva M, G Knight, T Hristov, D Mishev (eds) Global Change in Bulgaria, CIRA, the US and NCCGC-BAS, University Park and Sofia.

Part III
Environmental Impacts

Chapter 10
A Summary of Sector and Region Specific Economic Impact and Vulnerability Assessments by Case Study in Bulgaria, Romania and Hungary

Franz Prettenthaler and Judith Köberl

Abstract This paper summarises the main economic findings of an interdisciplinary EU project (FP6) which studied the impacts of climate change in Hungary, Bulgaria and Romania. Economic impact assessments were carried out at region and sector specific case study levels. Results suggest positive economic impacts related to climate change on agriculture in the Bulgarian case study region, whereas no clear message about impact direction could be derived for the Romanian region under investigation. The studied climate change impacts on tourism are tendentially negative in the regions considered within the single tourism case studies. The conducted energy case studies also suggest (small and sometimes insignificant) negative economic impacts of climate change within specific configurations assumed. Vulnerability of the public sector has been assessed by analysing the countries' risk transfer mechanisms with respect to damages from extreme weather events.

Keywords Climate Change · Economic Impact Assessment

10.1 Introduction

The nations in Central and Eastern Europe (CEE) face triple challenges, with ongoing economic and political transition, continuing vulnerability to environmental hazards, and long-term impacts of global climate change. The overall aim of EU FP6 project CLAVIER (CLimate ChAnge and Variability: Impact on Central and Eastern EuRope)[1] was to make a contribution to strategies in an attempt to

---

F. Prettenthaler (✉)
Joanneum Research, Institute of Technology and Regional Policy, Graz, Austria
e-mail: franz.prettenthaler@joanneum.at

[1]The CLAVIER project was supported by the European Commission's 6th Framework Programme (contract number 037013) as a 3-year Specific Targeted Research Project from 2006 to 2009 under the Thematic Sub-Priority "Global Change and Ecosystems". Daniela Jacob was coordinator of the project, Franz Prettenthaler led WP4 on the economic impacts of climate change.

V. Alexandrov et al. (eds.), *Global Environmental Change: Challenges to Science and Society in Southeastern Europe*, DOI 10.1007/978-90-481-8695-2_10,

successfully cope with these challenges. Three representative CEE countries were studied in detail: Hungary, Romania, and Bulgaria.

Within the framework of CLAVIER, ongoing and future climate changes were analysed in great detail based on existing data and climate projections to fulfil the need for a local and regional impact assessment. Researchers from six countries and different disciplines investigated links between climate change and its impact on weather patterns, air pollution, extreme events, and water resources. This paper summarizes some main findings of the final step of this interdisciplinary project, and contains an evaluation of economic impact on agriculture, tourism, energy supply and the public sector. Knowledge gained through achievement of scientific goals within meteorological, climatological and hydrological building blocks of these projects was the basis for our work as economists.

Economic impact assessments were undertaken at a region and sector specific case study level; based on climate simulations, potential physical impacts for selected case study regions were estimated and translated into economic impacts. Principally, the potential impacts of climate change were estimated by applying climate simulations resulting from the regional climate model REMO (Jacob 2001; Jacob et al., 2001; Jacob and Podzun 1997) for IPCC emission scenario A1B (as of now referred to as REMO-A1B). However, at least one case study investigating the same sector additionally conducted economic impact estimations based on simulations resulting from the regional climate model LMDZ (Hourdin et al., 2006) for IPCC emission scenarios A1B and B1 (as of now referred to as LMDZ-A1B and LMDZ-B1, respectively).

It is important to stress that the results of each case study below are only valid for the presented case study area and cannot be transferred to other regions. However, the methodology applied is transferable.

10.2 Agriculture

10.2.1 Bulgaria

The Bulgarian agriculture case study investigated potential economic impacts of climate change on agriculture in the North-East region. For this purpose, crop models for wheat, barley, maize and sunflowers were developed, whereby statistical relationship was calibrated by means of multiple linear regression using selected meteorological parameters as predictors and crop yields as predictands. Impact scenarios were developed on the basis of three different climate scenarios. Agricultural production figures resulting from average scenario yields for 2021–2030 were entered into a regionalized input-output table in order to derive the economic impact of investigated climate caused crop yield changes on total agricultural output as well as on the regional economy.

10.2.1.1 Main Results

As Figure 10.1 shows, results suggest that the average impact of climate change on crop yields during 2021–2030 will be positive or neutral under the three scenarios investigated with only one exception – maize under scenario REMO-A1B. The highest negative impact amounts to a yield reduction of 10% (maize under scenario REMO-A1B), the highest positive impact to a yield increase of 16% (barley under scenario LMDZ-A1B).

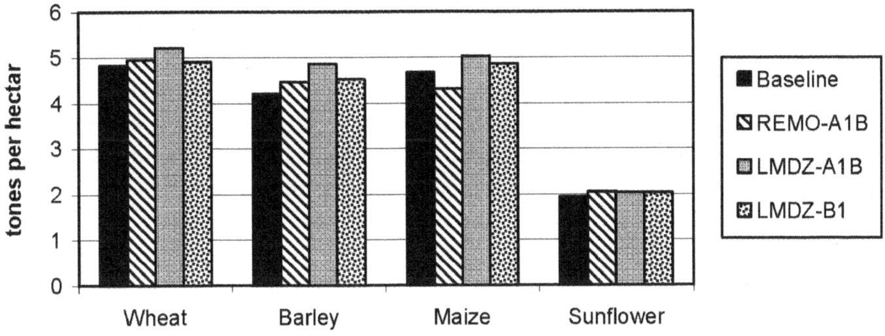

Fig. 10.1 Projected yields (tons per hectare) in the future 2021–2030. Note: Climate change impact is illustrated by differences in yields projected under climate scenario REMO-A1B, LMDZ-A1B or LMDZ-B1 compared to yields projected under the baseline scenario

The estimated economic impact of investigated climate caused changes in crop yields on gross agricultural output and total regional output are positive under all three scenarios (see Table 10.1). However, these favourable results should be regarded with some caution, as due to the inequity in the variation of some climate variables the positive impact of climate change could be overestimated.

Table 10.1 Economic impact of climate caused crop yield changes on gross agricultural output and total regional output in Bulgaria's North-East region

	Scenario			
	Baseline	REMO-A1B	LMDZ-A1B	LMDZ-B1
Gross agricultural output (mil. Leva)	2,164.40	2,915.00	3,219.99	3,110.67
Gross agricultural output (mil. €)	~ 1,340.87	~ 1,495.04	~ 1,651.46	~ 1,595.39
Difference to the baseline scenario (%)	–	+11.50	+23.16	+18.98
Total regional output (mil. Leva)	30,414.03	31,005.67	31,615.65	31,397.03
Total regional output (mil. €)	~ 15,598.67	~ 15,902.11	~ 16,214.95	~ 16,102.83
Difference to the baseline scenario (%)	–	+1.95	+3.95	+3.23

Main adaptation measures recommended for the North-East region include the introduction of drought-resistant crop varieties and moisture-saving technologies, improvements in data collection and provision at the regional scale, as well as improvements in the agricultural insurance system.

10.2.2 Romania

The Romanian agriculture case study focused on studying potential economic impacts of climate change on agriculture in the North-West region. For this purpose, crop models for wheat, barley, maize, potatoes, lucerne and clover were developed, whereby the statistical relationship was calibrated by means of multiple linear regression using selected meteorological parameters as predictors and crop yields as predictands. Impact scenarios were developed on the basis of three different climate scenarios. Agricultural production estimates for average scenario yields in 2020–2030 were applied to a regionalized input–output table in order to derive the economic impacts of investigated climate caused crop yield changes on total agricultural output as well as on the region's economy.

10.2.2.1 Main Results

As Figure 10.2 illustrates, results show that the average impact of climate change on single crop yields strongly depends on crop type as well as climate scenario.

The estimated economic impact of investigated climate caused changes in crop yields on gross agricultural output and total regional output is positive in the LMDZ-B1 scenario, negative in the REMO-A1B scenario and almost neutral in the LMDZ-A1B scenario (see Table 10.2).

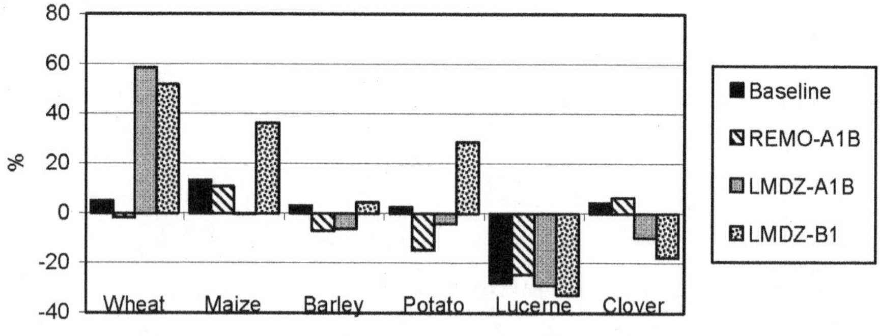

Fig. 10.2 Change in crop yields in the North-West region in 2020–2030 compared to reference period 1975–2000 according to different climate scenarios. Note: Climate change impact is illustrated by differences in yields projected under climate scenario REMO-A1B, LMDZ-A1B or LMDZ-B1 compared to yields projected under the baseline scenario

Table 10.2 Economic impacts of climate caused crop yield changes on gross agricultural output and total regional output in Romania's North-West region

	Scenario			
	Baseline	REMO-A1B	LMDZ-A1B	LMDZ-B1
Gross agricultural output (mil. lei)	13,192	12,401	13,137	14,963
Gross agricultural output (mil. €)	~3,669	~3,449	~3,653	~4,161
Difference to the baseline scenario (%)	–	−6.00	−0.42	+13.42
Total regional output (mil. lei)	116,624	115,833	116,569	118,395
Total regional output (mil. €)	~32,434	~32,214	~32,419	~32,926
Difference to the baseline scenario (%)	–	−0.68	−0.05	+1.52

Due to the differences between climate models for future yield projections and because of territorial differences within case study area borders, it is difficult to draw a uniform conclusion with respect to an adaptation strategy.

10.2.3 Comparative Conclusion

Whereas the economic impact of investigated climate caused changes in crop yields on the region's gross agricultural output as well as on the whole region's economy is positive in the case of the Bulgarian North-East region under all three applied climate scenarios, economic impacts in the case of the Romanian North-West region Romania are differential, depending on the climate scenario applied. Thus, uncertainty is too high to say whether the agricultural sector and regional economy of Romanian North-West region will be positively or negatively influenced by climate change induced crop yield differences.

10.3 Tourism

10.3.1 Bulgaria (Winter Tourism)

This section summarizes the results of a case study also presented in more detail in another chapter of this book (see Mishev, Mochurova). The Bulgarian tourism case study aimed at investigating the potential economic impacts of climate change on winter tourism for the ski resort Borovets, located in Samokov. By means of regression analysis, the relationship between tourism performance (in terms of overnights in Borovets) during the winter season (October–March) and meteorological parameters (the number of snow days and snow height) was studied. Impact scenarios were based on climate simulations from the regional climate model REMO 5.7 for the A1B emission scenario.

10.3.1.1 Main Results

The economic impact of a changing climate on winter tourism in Borovets, as quantified by the regression analysis carried out, is negative, but almost negligible. The reasons for this outcome are 2-fold:

- Although there is a stable downward trend in the number of snow days and in snow height, even in the period 2021–2050 ski zones will on average still have enough snow for skiing, according to the REMO-A1B scenario.
- It was not possible to quantify the effects of more intensive cyclic development in snow cover, which is expected to lead to a higher frequency in snow-deficient years for the period 2021–2050. However, with existing patterns of planning ski vacancies in advance this could cause a real sharp decline in the number of tourists and thus is of economic importance for the resort.

Short-term adaptation measures relate to the economic agents of the resort, which should engage in diversification of services offered to tourists. Long-term measures involve the municipalities, which should try to diversify economic activities and not rely so extensively on winter tourism.

10.3.2 Hungary (Summer Tourism)

The Hungarian tourism case study focused on analysing potential economic impacts of climate change on summer tourism at Lake Balaton and in Veszprém, respectively. In order to analyze the relationship between tourism performance and climate during the main season (July and August), multiple regression analyses were performed using selected meteorological parameters (amongst others, mean air temperature) as predictors and tourist arrivals as predictand. Impact scenarios were developed for the periods 2016–2025 and 2041–2050 on the basis of three different climate scenarios.

10.3.2.1 Main Results

Depending on the climate scenario, the impact of changing monthly mean air temperatures on tourist arrivals and expenditures vary slightly in their extent, but show the same trend. In the case of Lake Balaton, regression results suggest weak positive effects for the periods 2016–2025 and 2041–2050 compared to the reference period 1995–2006, indicating climate will cause an increases in tourist expenditures of about 1% to 2%. For Veszprém, however, regression results show slightly negative effects for the two investigated future periods compared to the reference period, suggesting climate caused decreases in tourist expenditures of up to about –2%. However, it is important to stress that effects of extreme events have not been considered with the applied method.

Additional qualitative impact assessment shows that the number of days at which some portion of the population feels discomfort (measured by the Thom's

Discomfort Index[2]) is expected to increase during the main season (July and August) according to the REMO-A1B scenario. However, there are some aspects mitigating the problem of rising uncomfortable days:

- The northern Adriatic Sea, which Lake Balaton directly competes with, is predicted to face similar problems.
- Lake tourism in the region under investigation could benefit from the fact that the increase in the number of uncomfortable days is expected to be higher in Budapest than at Lake Balaton.
- Rising air temperatures are expected to lead to a longer peak season and increasing numbers of trips outside the usual season.

Thus, by using the partially positive effects of climate change, the tourism industry in the region could even benefit from climate change.

10.3.3 Romania (Summer Tourism)

The Romanian summer tourism case study investigated the potential economic impacts of climate change on summer tourism at the Black Sea Coast (Constanta County). In order to analyze the relationship between tourism performance and climate at the Black Sea coast during the main season (July and August), regression analysis was performed using mean air temperature (and its squared term) as predictors and the number of overnight stays as predictand. The impact scenario was developed on the basis of climate scenario REMO-A1B. Using input-output analysis, potential impacts of investigated climate-caused changes in overnights at the Black Sea coast on the national economy were assessed.

10.3.3.1 Main Results

Input-output analysis suggests that a reduction in tourism demand at the Romanian Black Sea coast, as indicated by regression analysis due to an increase in mean air temperature as predicted by the REMO-A1B scenario, will lead to a production loss at the national level of 0.7% in 2020 compared to 2005. This result highlights the importance of regional adaptation measures.

10.3.4 Romania (Winter Tourism)

The Romanian winter tourism case study focused on investigating the potential economic impacts of climate change on winter tourism at the ski resorts

[2]The Thom's Discomfort Index (THI) includes the parameters air temperature and relative humidity (see e.g. Tzenkova et al., 2003).

Sinaia (Prahova) and Predeal (Braşov). To analyse the relationship between tourism performance and climate during the winter season (November–April) at the mentioned resorts, regression analyses were carried out using selected meteorological parameters as predictors and selected tourism indicators as predictands. An impact scenario was developed for the periods 2020–2025 and 2045–2050 on the basis of the REMO-A1B climate scenario.

10.3.4.1 Main Results

In the case of Sinaia, regression results suggest that changing mean air temperatures as simulated by the REMO-A1B scenario for the periods 2020–2025 and 2045–2050 compared to the reference period 2002–2007 will cause a decrease in tourist expenditures of about 0.5 and 1.1%, respectively. A more negative impact of changing mean air temperatures is indicated for Predeal, where climate change as simulated by the REMO-A1B scenario is expected to cause reductions in tourist expenditures of about 4 and 9%, respectively. However, it is important to stress that effects of extreme events have not been considered within the applied method.

In order to minimize the negative impacts of climate change on the tourism sector in the area under investigation, it is recommended to integrate climate change issues into local and regional tourism policies as a tangible point and at the same time to diversify recreational offers in general.

10.3.5 Comparative Conclusion

The results of single case studies suggest that climate change tendentially impacts tourism in the regions under consideration in a negative way.

10.4 Energy Supply

10.4.1 Bulgaria (Nuclear Power)

The Bulgarian energy case study investigated the potential economic impact of climate change on nuclear power production at the Kozloduy Nuclear Power Plant (KNPP), located in the North-West region. In order to assess potential economic impacts of climate change as simulated by three different climate scenarios, comparative analysis, scenario analysis and descriptive statistical methods were applied. Since ordinary (normal) climate-related conditions do not have any influence on electricity production and do not affect the safety of nuclear units and nuclear waste, analysis focused on the impact of extreme weather events.

10.4.1.1 Main Results

Investigation of maximum and minimum daily temperatures, the heat wave duration index, maximum precipitation and the precipitation intensity index as projected in the three different scenarios for the period 2021–2050 leads to the conclusion that projected climate change will not influence the safety of the KNPP, as expected extreme events are within design capacity limits for the power plant. However, climate change is expected to diminish cooling efficiency during summer months and hence lead to decreased energy production. More precisely, the number of unfavourable days per year (days with low cooling efficiency at KNPP, i.e. $T > 30°C$) is projected to increase between 1.6 and 5 times in the period 2021–2050 when compared to the period 1961–1990 according to the three climate scenarios (see Table 10.3). Nevertheless, expected physical and economic impacts of climate change on nuclear energy production are low in all scenarios. Taking only unfavourable days during the summer months (June–August) into account, yearly economic losses of KNPP due to climate change are expected to range from about € 5.4 million to € 7.4 million (see Table 10.4). The expected decrease in electricity production only accounts for a small portion of KNPP's total production in 2007, namely about 1%. Hence, expected economic impacts are insignificant at both regional and national levels.

Regarding adaptation, better planning with respect to regular annual maintenance – for which a unit has to be shut down – is recommended by choosing the hottest periods of the year for this activity, when cooling efficiency is low anyway.

Table 10.3 Average number of days per year with T > 30°C

	REMO-A1B	LMDZ-A1B-L	LMDZ-B1-L
1961–1990	33	14	14
2021–2050	52	53	72
Difference	+ 58%	+ 279%	+ 414%

Table 10.4 Expected yearly losses at KNPP caused by climate change in 2021–2050

		Decreased electricity production		Economic losses	
Scenario	Average number of days per year with T > 30°C in summer months (days)	Absolute (GW.h)	Share of total production 2007 (%)	(in 1,000 BGN)	(in 1,000 €)
REMO-A1B	43	100.4	0.7	10,568.26	~5,404
LMDZ-B1	59	137.7	1	14,500.64	~7,414

10.4.2 Hungary (Wind Energy)

The Hungarian energy case study focused on investigating the potential economic impacts of climate change on the energy production of a wind farm in Győr-Moson-Sopron. To this end, annual energy yields were calculated based on statistical relations between modelled daily wind speeds and observed power outputs (10 min mean values) of the wind turbines, taking into account annual turbine failure and wake-losses of the wind farm. Furthermore, the effects of climate change on air density and wake-losses were investigated. Impact assessment was carried out on the basis of the REMO-A1B scenario.

10.4.2.1 Main Results

Comparing 10-year averages of scenario period 2021–2030 and reference period 2003–2012, mean annual energy yields are expected to be insignificantly reduced by −1.5% due to changing wind speeds. However, significant (significance <0.05) changes are expected to occur later in the future (Figure 10.3, left panel). The trend of air density for the scenario and reference period is small (−0.03%) and statistically insignificant. Minor (between ~0.0 and −1.0%), but statistically significant changes (based on 10-year moving averages) may arise after 2028 (Figure 10.3, right panel). The wake losses of the wind farm show complex behaviour: they are supposed to depend on both direction and velocity of approaching air flows. Nevertheless, since regional climate simulations (REMO-A1B) do not indicate climate change effects on wind direction, and since expected changes of wind speeds are insignificantly small between the scenario and the reference period, the impact on energy yield should also be negligible within the given configuration. In conclusion, the potential economic impacts on wind farming in Győr-Moson-Sopron are expected to be tendentially negligible.

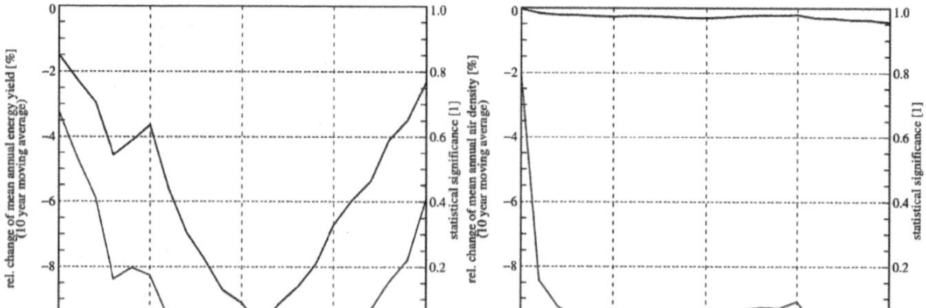

Fig. 10.3 Climate change effects (*black*) (relative difference to reference period 2003–2012) and their significance (*grey*). Note: Changes in the wind farm's mean annual energy yield calculated from daily wind speeds and current observed technical availability of wind turbines (*left panel*) and changes in mean annual air density (*right panel*) are shown

An analysis of the region's adaptive capacity shows a relatively high adaptation potential for Győr-Moson-Sopron: adaptation is affordable both in terms of tangible and intangible assets.

10.4.3 Romania (Hydropower)

The Romanian energy case study focused on investigating the potential economic impacts of climate change on hydropower production at the Vidraru Hydropower Plant (part of the Hidroelectrica Company) and on evaluating the regional vulnerability of Argeş County (NUTS III), where the plant is located. An analysis of potential climate change impacts on hydropower production was carried out by investigating the system's behaviour within the reference period (1961–1990) and using the obtained results as well as climate change predictions to develop impact scenarios for a future period (2021–2050). Simulations of future climatic and hydrological conditions were based on REMO-A1B simulations and the hydrological model CONSUL.[3]

10.4.3.1 Main Results

Based on the relationship between measured energy and inflow data at Vidraru Reservoir (determined by means of regression analysis) and inflow projections according to the A1B scenario, it is estimated that Vidraru hydropower plant annual electricity production will decrease slightly in 2021–2050 compared to the reference period 1961–1990 (< 5% decrease in annual energy production). However, as the share of the Vidraru hydropower plant in the annual electricity output of the Hidroelectrica Company lies below 3%, the national significance of climate change caused impacts on electricity production at the Vidraru hydropower plant is small. Nevertheless, climate change should be embodied as a priority in policy making for the hydropower sector in order to realise integrated adaptation, considering major climatic threats.

10.4.4 Comparative Conclusion

All three energy case studies suggest small (and sometimes insignificant) negative economic impacts of climate change within the specific configurations assumed in each case study (regarding the region under investigation, the applied climate scenario, the type of electricity production, the period investigated and the meteorological parameters considered).

[3]The hydrological model was developed by the Romanian National Institute of Hydrology and Water Management.

10.5 Public Sector

A well-implemented plan describing how to spread economic risks from extreme events within society and/or transfer them from victims to financial markets is a fundamental adaptation measure that will crucially decide how disturbing impacts from climate change will be for society. By examining the institutional setting of risk transfer in Bulgaria, Hungary and Romania, and looking into the funds used to deal with such emergencies, the public sector case study aimed at gaining insights into public sector vulnerability to extreme events. Contrary to the other three sectors investigated, the public sector's vulnerability was studied at a national level. Improvement potentials were identified based on an analysis of the desired elements for an efficient risk transfer mechanism (Prettenthaler and Albrecher 2009).

10.5.1 Bulgaria

In Bulgaria, insurance against natural hazards is voluntary. Due to long and complicated procedures, little confidence exists in the private insurance market. Thus, catastrophe insurance penetration regarding natural hazards is quite low. Only about 7% of Bulgarian homeowners are insured against damages arising from natural disasters (Gurenko et al., 2008). Because of low insurance penetration, the Bulgarian government has provided financial assistance to uninsured homeowners following floods in recent times. However, this is a strategy the Bulgarian state cannot afford, since financial preparedness to cope with major disasters is quite low. Thus, several proposals involving reforming the risk transfer scheme have appeared recently, ranging from the installation of a national catastrophe insurance pool similar to the Romanian or Turkish scheme, to participation in the regional catastrophe insurance program the World Bank is preparing for South Eastern (and Central) Europe.

Major improvement potentials in the current scheme

- Establishment of a (well-coordinated) risk partnership between insurers, state and citizens
- Generation of a bigger risk collective (e.g. through the introduction of mandatory elements)
- Establishment of explicit rules for the handling of objects with very high damage frequency

10.5.2 Hungary

In Hungary, insurance penetration against floods, which represent the natural hazard the country is mostly at risk of, is quite high. However, coverage offered by private insurers is extremely limited and people living in poor areas that are highly exposed to floods often have no access to private insurance. Since 2003, a state fund has existed for homes situated in highly exposed flood plains, which have difficulty getting private insurance coverage.[4] Homeowners whose dwellings are located in risky regions may contribute to the fund. In turn, they are entitled to indemnification in case of damage. However, as the fund is mainly financed by the state budget, public sector vulnerability remains.

Major improvement potentials in the current scheme

- Establishment of a (well-coordinated) risk partnership between insurers, state and citizens
- Creation of socially acceptable premiums for 100% coverage (premium rates of state-offered flood insurance increase with decreasing value of buildings and seem rather expensive for small dwellings)

10.5.3 Romania

In Romania, a concrete reform process regarding a financial risk transfer mechanism with respect to natural hazards is in progress at the moment. Work is being done to establish a national catastrophe insurance pool founded through the affiliation of insurance companies. Moreover, household insurance against certain natural risks, namely earthquakes, floods and landslides, is expected to become mandatory by 1 January 2010. The whole system of mandatory household insurance will be guaranteed by the state, acting as a reinsurer of last resort. At the moment there are still unresolved problems with respect to the new system. However, if these problems are successfully solved, the new system is likely to increase adaptation capacity considerably and thus decrease vulnerability of the public sector and society as a whole; at the moment household insurance is still voluntary, and only 11–12% of the 8 million homes in Romania are insured (Cracea 2009).

Major improvement potentials in the planned scheme

- Better incentives for individual risk prevention (e.g. through the introduction of risk-based premiums and deductibles)

Acknowledgment The authors wish to thank other authors of individual case studies for their contributions: Plamen Mishev, Antonaneta Golemanova, Nedka Ivanova, Milkana Mochurova and Todor Kaloyanov for the Bulgarian case studies, Dan Balteanu, Ana-Irina Dinca, Diana Dogaru, Carmen Dragotă, Bianca Dumitrescu, Monica Dumitraşcu, Dana and Mihai Micu, Camelia Surugiu, Marius Matreata, as well as Boróka Bíró, Emese Szocs and Maria Vincze for the Romanian case studies and Eric Kirschner, Heimo Truhetz, Andreas Gobiet, Cornelia Sterner and Petra Amrusch for the Hungarian case studies. For detailed results see Jacob (ed.), 2010 and www.clavier-eu.org.

[4] See Law 58/2003 regarding the Wesselényi Miklós Flood and Inland Water Compensation Fund.

References

Cracea M (2009) The AOL Law may suffer further modifications, Article published in XPRIMM Newsletters on 25th June 2009.

Gurenko EN, Itigin A, Dumitru D (2008) Bulgarian Catastrophe Insurance Initiative: Feasibility Study, Project Report, World Bank.

Hourdin F, Musat I, Bony S et al. (2006) The LMDZ4 general circulation model: Climate performance and sensitivity to parametrized physics with emphasis on tropical convection. Clim Dyn 27:787–813.

Jacob D (2001) A note to the simulation of the annual and inter-annual variability of the water budget over the Baltic Sea drainage basin. Meteorol Atmos Phys 77:61–73.

Jacob D (ed) (2010) Climate Change in Hungary, Romania and Bulgaria: Variability and Impact, Imperial College Press, London, Forthcoming.

Jacob D, Podzun R (1997) Sensitivity studies with the regional climate model REMO. Meteorol Atmos Phys 63:119–129.

Jacob D, van den Hurk BJJM, Andrae U et al. (2001) A comprehensive model inter-comparison study investigating the water budget during the BALTEX-PIDCAP period. Meteorol Atmos Phys 77:19–43.

Prettenthaler F, Albrecher H (eds) (2009) Hochwasser und dessen Versicherung in Österreich, Verlag der Österreichischen Akademie der Wissenschaften, Wien.

Tzenkova A, Kandjov I, Ivancheva J (2003) Some biometeorological aspects of urban climate in Sofia. Paper presented at the Fifth International Conference on Urban Climate, Lodz, Poland, 1–5 September 2003

Chapter 11
Cecilia – EC FP6 Project on the Assessment of Climate Change Impacts in Central and Eastern Europe

Tomas Halenka

Abstract The presented EC FP6 Project CECILIA is studying climate change impacts in Central and Eastern Europe based on the dynamical downscaling, i.e., nesting of a fine scale limited area model (or Regional Climate Model, RCM) within the Global Climate Models (GCMs). Global Climate Models (GCMs) can reproduce climate features on large scales, but their accuracy decreases when proceeding from continental to regional and local scales because of the lack of resolution. This is especially true for surface fields, such as precipitation, surface air temperature and their extremes, which are critically affected by topography and land use. However, in many applications, particularly related to the assessment of climate change impacts, the information on surface climate change at regional to local scale is fundamental. In the region of Central and Eastern Europe the need for high resolution studies is particularly important. A resolution sufficient to capture the effects of these topographical and associated land-use features is necessary. Therefore, 10 km resolution has been introduced in the presented project CECILIA for regional climate modelling in targeted areas of Central and Eastern Europe to provide satisfactory resolution in local impact studies covering key sectors of the region like hydrology, water quality, and water management, air quality issues, agriculture and forestry.

Keywords Climate change · Climate change impacts · Dynamical downscaling · Regional climate modelling

---

T. Halenka (✉)
Department of Meteorology and Environment Protection, Charles University, Prague, Czech Republic
e-mail: tomas.halenka@mff.cuni.cz

V. Alexandrov et al. (eds.), *Global Environmental Change: Challenges to Science and Society in Southeastern Europe*, DOI 10.1007/978-90-481-8695-2_11,

11.1 Introduction

Although the broad response of global climate to increased greenhouse gas concentrations is well established, many unknowns remain in the regional details of projections of future climate change. The floods and droughts which occurred in recent summers in the region highlight the importance of the hydrologic cycle and water management in Elbe and Danube river catchments in response to the occurrence of precipitation extremes. In summer of 2002 the Czech Republic experienced some of its worst floods in history with the Vltava river inundating Prague causing severe and widespread damage. The 2003 heat wave, one of the severest heat waves on record in central and western Europe causing both human losses (Kysely 2004) and extensive damage to human activities and natural ecosystems, demonstrated the importance of the health impacts of extreme conditions that could also lead to considerable changes in air quality, both regionally and in major urban centres. The possibility of changes in the frequency and intensity of these extreme events is one of the most dreaded manifestations of anthropogenic climate change. A number of studies have linked the occurrence of these extreme events to anthropogenic forcings (Beniston and Stephenson 2004; Schär et al. 2004; Pal et al. 2004; Meehl et al. 2004; Meehl and Tebaldi 2004). Impacts on agriculture and forestry affecting the economy of countries in the region are extensively studied as well (Menzel et al. 2006, 2003, Hafner 2003, Gobron et al. 2005).

While coupled atmosphere-ocean general circulation models (AOGCMs) provide basic information on the development of the climate change, their horizontal resolution is still too coarse to describe in detail processes affecting extreme events at the regional scale (Gates et al. 1996). In order to regionally enhance the AOGCM information a number of regionalization techniques have been developed (Giorgi et al. 2001). One of them is the use of limited area regional climate models (RCMs) nested in driving fields from reanalysis or GCMs similarly as previously done in numerical weather prediction for decades (Giorgi and Mearns 1999). The main advantage of this method is that it can reach with the same or even less resources horizontal grid intervals of a few tens of km and thus RCMs can provide improved simulation of extreme events (e.g. Huntingford et al. 2003; Frei et al. 2003). During the last decade RCMs have been increasingly used to examine climate variations at scales that are not resolved by global models. For Central Europe domain after the floods in 2002 Halenka et al. (2006) tested the ability of RCM to reproduce precipitation extremes. To the extent that they produce realistic climate simulations, such models can be powerful tools in the study of regional climate impacts.

Thus, the aim of the EC FP6 project CECILIA (**C**entral and **E**astern Europe **C**limate Change **I**mpact and Vulnerabi**LI**ty **A**ssessment), starting on June, 1, 2006, and extended till 31 December 2009 (for list of partners see Table 11.1), is to assess the impact of climate change at the regional to local scale for Central and Eastern Europe (CEE).using very high resolution simulations in order to capture the effects of the complex terrain of the region. In this region of Central and Eastern Europe the need for very high resolution studies is particularly important. This region is characterized by the northern flanks of the Alps, the long arc of the Carpathians, and

Table 11.1 List of partners

Participant name	Participant short name
Charles University, Prague, Czech Republic	CUNI
The Abdus Salam ICTP, Trieste, Italy	ICTP
Météo-France, Toulouse, France	CNRM
Danish Meteorological Institute, Copenhagen, Denmark	DMI
Aristotle University of Thessaloniki, Greece	AUTH
Czech Hydrometeorological Institute, Prague, Czech Republic	CHMI
Institute of Atmospheric Physics, Prague, Czech Republic	IAP
Swiss Federal Institute of Technology Zurich, Switzerland	ETH
University of Natural Resources and Applied Life Sciences, Vienna, Austria	BOKU
National Meteorological Administration, Bucharest, Romania	NMA
National Institute of Meteorology and Hydrology, Sofia, Bulgaria	NIMH
National Institute of Hydrology and Water Management, Bucharest, Romania	NIHWM
Hungarian Meteorological Service, Budapest, Hungary	OMSZ
Forest Research Institute, Zvolen, Slovakia	FRI
Warsaw University of Technology, Warsaw, Poland	WUT
Eötvös Loránd University, Budapest, Hungary	ELU

smaller mountain chains and highlands in the Czech Republic, Slovakia, Romania and Bulgaria that significantly affect the local climate conditions. A resolution sufficient to capture the effects of such topographical and associated land-use features is necessary as illustrated in Fig. 11.1, where comparison of model topography representation in resolutions of 50 and 25 km used for EC FP6 project ENSEMBLES to the resolution of 10 km introduced in CECILIA project is presented in the detailed view on the Czech Republic. Additionally, the central internal objectives of CECILIA are to improve regional climate scenarios and their localization for climate impacts models, and comparing these results against the results of previous and ongoing projects to assess the added value of dynamical downscaling at very fine scales (Table 11.1).

These goals are achieved using very high resolution RCMs run locally for targeted areas. Changes in weather patterns and extreme events are addressed within the project as they affect the important sectors for the economies and welfare of individual countries in the region. Uncertainties are evaluated by comparing results with those from previous projects (PRUDENCE, ENSEMBLES). The selected applications of the CECILIA outputs are supposed toward water resources and management, agriculture, forestry, air quality and health. In addition, CECILIA will improve the access of CEE researchers to information and facilities for climate

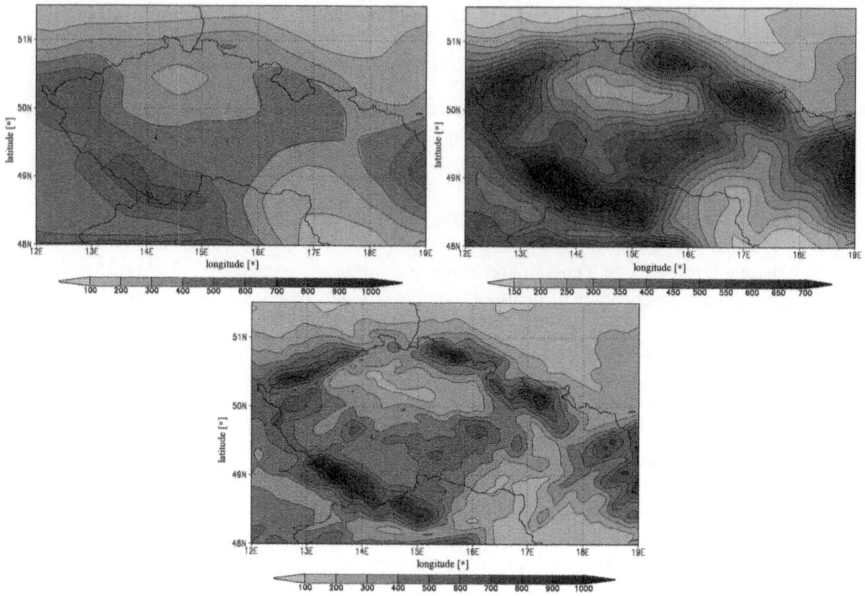

Fig. 11.1 Detail of topographical features seen in ENSEMBLES' 50 km resolution (*upper left*) and 25 km (*upper right*) and 10 km for CECILIA proposal (*bottom panel*)

change research by providing an efficient use and access to the results of previous and ongoing EC projects which the proposed research will benefit greatly from, e.g.:

- "Modelling the Impact of Climate Extremes (MICE)"
- "Statistical and regional dynamical downscaling of extremes for European regions (STARDEX)".
- "Prediction of Regional scenarios and Uncertainties for Defining EuropeaN Climate change risks and Effects" (PRUDENCE)
- "ENSEMBLE-based Predictions of Climate Changes and their Impacts" (ENSEMBLES)
- "Quantifying the Climate Impact of Global and European Transport Systems" (QUANTIFY).

Thus, CECILIA is integrating world leading European expertise in regional climate modelling with high resolution impact studies to provide new policy relevant information on climate change and its interactions with society at the regional scale. It will also feed into adaptation and mitigation strategies in targeted areas.

11.2 Key Issues

Emphasis in CECILIA project is given to application of regional climate modelling studies at a resolution of 10 km for local impact studies in key sectors of the region. From the viewpoint of climate change scenario production two time slices are planned, for 2020–2050 and 2070–2100. The high spatial and temporal resolution of national observational networks and of regional model experiments will feed into investigations of consequences for weather extremes in the region. Statistical downscaling methods for verification of the regional model results will be developed and applied, and assessments of their use in localization of model output for impact studies will be performed. The objectives will be achieved through the following tasks:

- To collect, assess and make available for first local impact studies the scenarios and climate simulations produced in previous relevant projects where available (WP1).
- To adapt and develop very high resolution RCMs for the region (10 km grid spacing) and perform regional time-slice nested runs driven by ERA40 data and by GCMs for selected GHG change scenarios (WP2).
- To verify the model results, compare RCM and statistical downscaling results, analyze and develop the methods for verification, particularly at local scales, to provide the scenarios (WP3).
- To estimate the effect of global climate change on extreme events in the region, including the assessment of the added value of high-resolution for the simulation

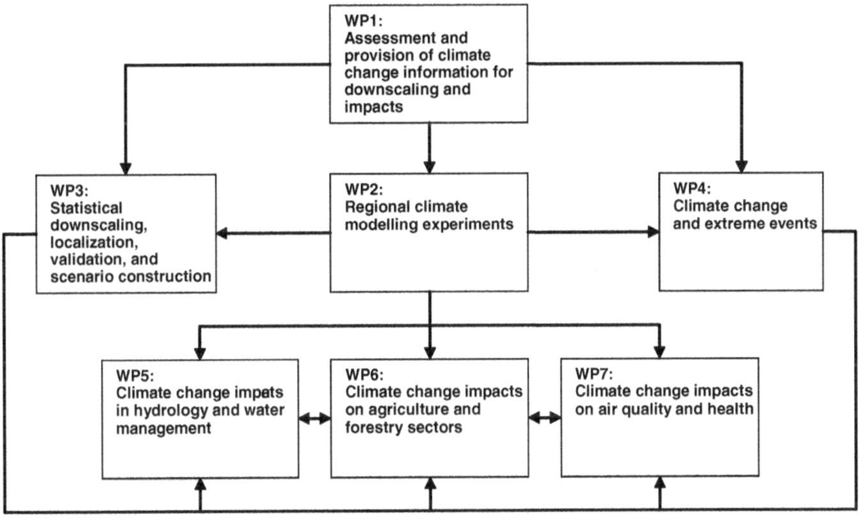

Fig. 11.2 Interactions between the workpackages

of the relevant processes and feedbacks. To evaluate uncertainties in regional projections by comparing results from previous projects (WP4).

- To assess (using high resolution downscaling results) the impacts of climate change on the hydrological cycle and water resources over selected catchments; the effects of climate change on the Black Sea (WP5).
- To study (based on the high resolution downscaling results) the impacts of climate change on agriculture and forestry, carbon cycle and selected species (WP6).
- To study (based on the high resolution downscaling results) the impacts of climate change on health and air quality (photochemistry of air pollution, aerosols) (WP7).

The overall structure and dependencies within the project are given in Fig. 11.2.

11.3 Expected Impact

Although the response of global climate to increased greenhouse gas concentrations is well established, many unknowns remain in the regional details of projections of future climate change. Thus, the central objectives of CECILIA are to improve regional climate scenarios and their localization for climate impacts models, and comparing these results against the results of previous and ongoing projects to assess the added value of dynamical downscaling at very fine scales. Several key issues connected with climate change have become of interest in recent years, such as the occurrence of extremes or effects on air quality, with potentially severe impacts on the quality of life, health and safety. The occurrence of these extreme events, in some cases causing loss of human life and extensive damages or costs, is affected by the relation between extremes and climate change which can be better explored using high resolution climate modelling. Results will allow us to evaluate the vulnerability of different sectors in the regions. CECILIA will help to identify and exploit positive impacts. It will provide demonstrations of the use of these tools in important economic, environmental or social sectors where the impacts of climate change are likely to be felt.

Climate change represents a major factor affecting the global and European environments. Natural ecosystems will become stressed if climatic zones shift at a faster rate than the ecosystems can migrate. Changing availability of natural resources such as water supply may adversely affect the sustainability of European activities. A more stressed environment will be even more vulnerable to natural hazards. CECILIA with high resolution climate simulation can help anticipate and ameliorate the adverse impacts on the local environment and natural resources of the targeted regions. It can also provide mitigation information to reduce the hazards concerning these important factors. Concerning the environment, CECILIA, similarly as the EC project QUANTIFY, will provide a platform for reducing the gap between climate change and air quality sciences, putting together traditional aspects of climate change impacts and impacts on air quality.

This project brings very high resolution localization of climate change scenarios into the targeted areas of CEE, with the added value of climate scenarios produced locally. This will provide necessary policy relevant information concerning the local adaptation and/or mitigation measures. Moreover, it will provide know-how and tools which can be further used for the analysis of the climate change development and climate change impacts on different sectors of the society in the target region.

11.4 Preliminary Results

In framework of a review of previous available results (PRUDENCE project) first estimates of climate change signal for targeted regions were publicised and presented to end-users (see Fig. 11.3). More detailed analysis as well as the comparison with other results (IPCC 4AR model results) are under preparation for publication. Analysis in Fig. 11.4 reveals that the data from recent database of CMIP3 ensemble exhibits significant warming in northeastern Europe during winter whereas in Mediterranean in summer, for precipitation the area of no significant changes is moving across the Central Europe depending on seasons with significant drying in Mediterranean in summer and increase of precipitation in northern Europe during winter.

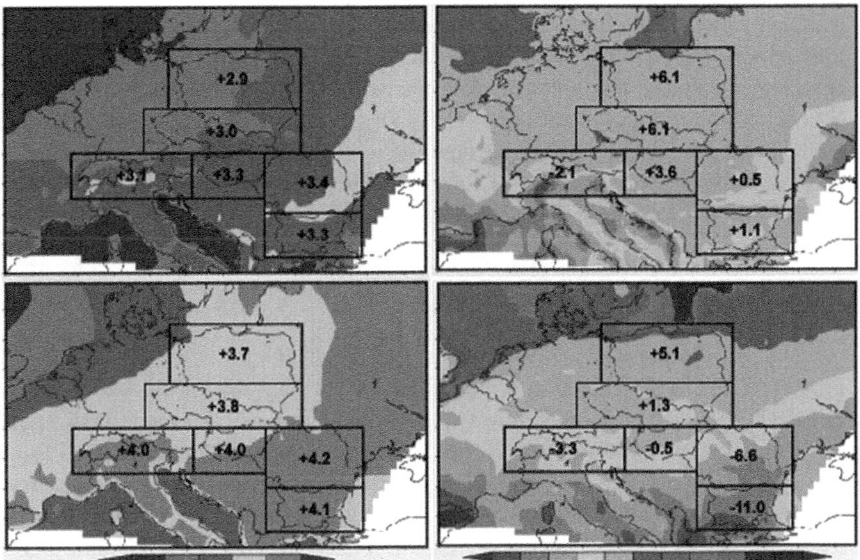

Fig. 11.3 Change of temperature (°C – *left column*) and precipitation (% – *right column*) in 2071–2100 relative to 1961–1990 in the CECILIA target areas for emission scenarios A2 (*bottom panels*) and B2 (*upper panels*). (ICTP provided analysis)

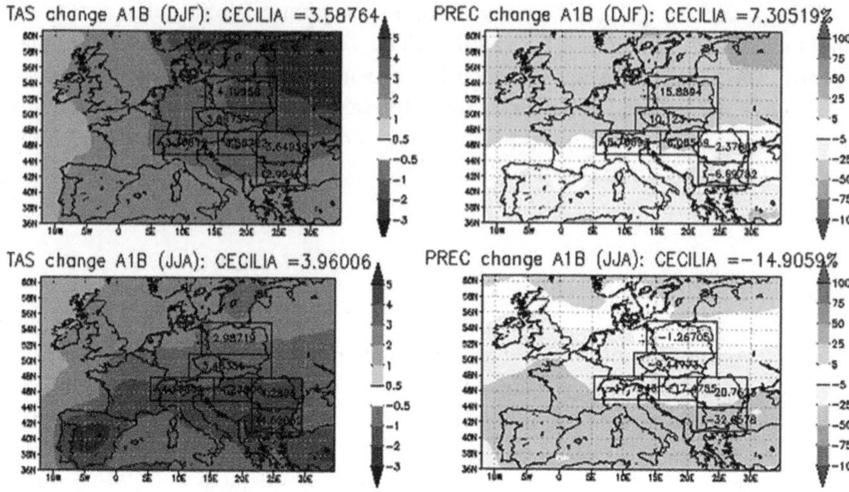

Fig. 11.4 Change of temperature (°C – *left column*) and precipitation (% – *right column*) in 2071–2100 relative to 1961–1990 in the CECILIA target areas for emission scenario A1B in winter (*upper panels*) and in summer (*bottom panels*) for the CMIP3 global models (IPCC 4AR). (ICTP provided analysis)

Although used as the first stream of data, the coarse temporal and spatial resolution makes CMIP3 difficult for use in impact part of our project. The best tool to start and validate the impact studies is the FP5-PRUDENCE database in daily resolution and in 50 km, which is closer to what needed for CECILIA, in Fig. 11.5 it can be seen reasonable consistency of these data with the new results. Averages for each country of the region (according to the political boundaries at 50 km resolution) have been calculated for targeted countries with model responses for each model and SRES-A2 scenario, in Fig. 11.5 provided for temperature and precipitation as an example for Bulgaria.

11.5 CECILIA Experiments

For CECILIA high resolution simulations six regional climate models covering Central Europe (Czech Republic – two models), Hungary (two models), Romania and Bulgaria have been prepared for present day climate simulation as well as climate change scenarios in a high horizontal resolution (10 km) on the domains shown in Fig. 11.6. The main objective is to produce simulations on targeted domains for a past period (1961–1990) driven by ERA40 reanalysis, for a reference period (1961–1990) and two GCM driven scenario time slices (2021–2050 and 2071–2100) based on AR4-A1B GCM projections. CUNI, ELU and NMA are using RegCM (Pal et al. 2007; Elguindi et al. 2006), CHMI, OMSZ and NIMH are using ALADIN-Climate (Huth et al. 2003). In Fig. 11.7 the climate change signal under scenario

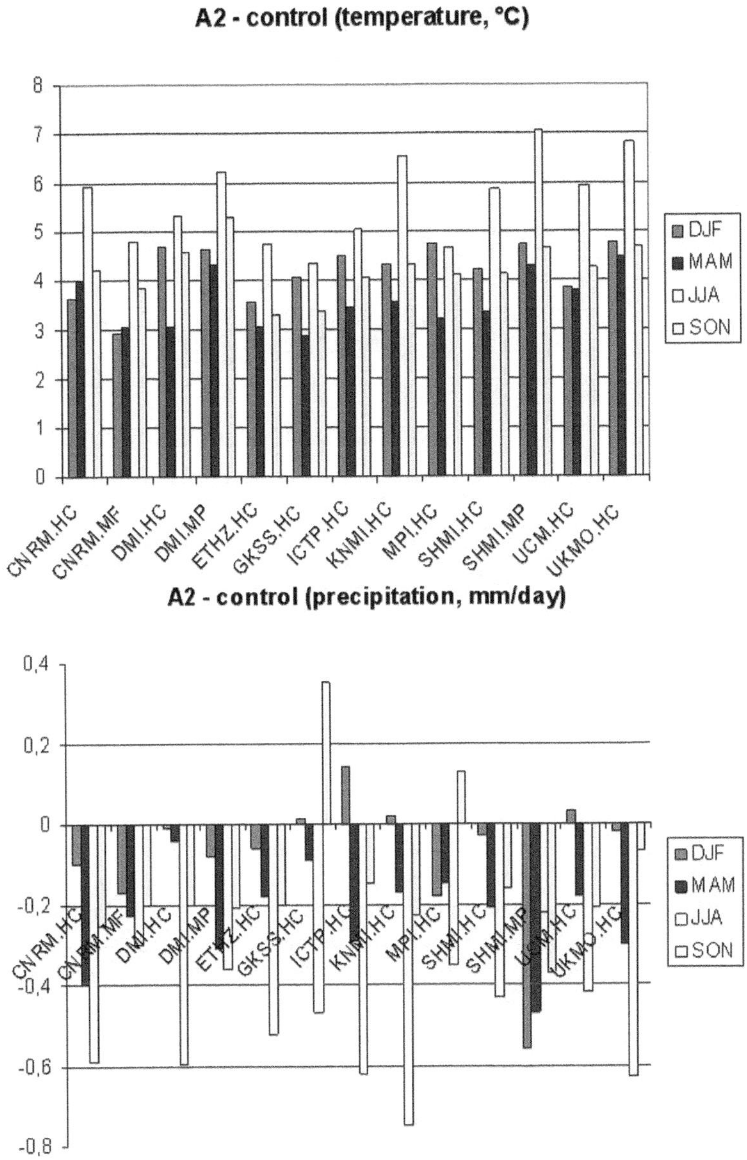

Fig. 11.5 Temperature response (°C) and precipitation response (mm/day) of the PRUDENCE A2 simulations over the Bulgaria for four seasons. (Based on Meteo France results)

A1B for surface temperature is presented as an example from CUNI simulations for Central Europe, driven by ICTP RegCM@25 km run for EC FP6 ENSEMBLES Project (forced by ECHAM5 run for A1B scenario). The warming of about 1°C and about 3°C clearly seen in Central Europe for mid-century time slice and end-of-century time slice, respectively, is consistent with the results of IPCC AR4 GCMs,

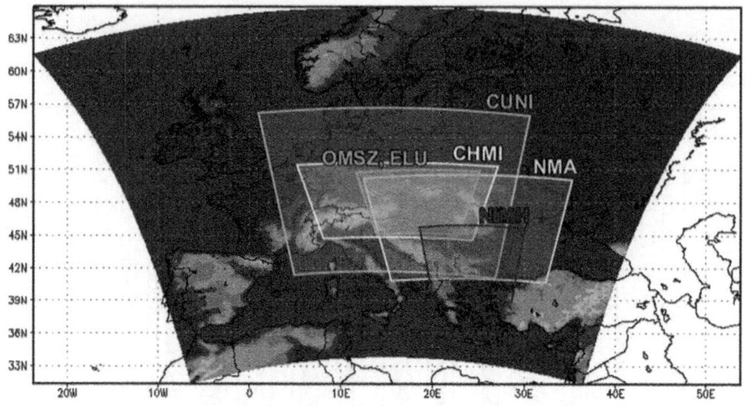

Fig. 11.6 Domains of simulations for individual partners in targeted regions

the annual patterns formed by increased warming to the north and east during winter and spring seasons while enhanced warming towards the Mediterranean and Eastern Europe contributing mainly in summer.

11.6 Conclusions

The general aim of CECILIA is to improve Europe's ability to assess the consequences of global climate change at the local scale, and on this basis to assist to formulate more precise response strategies and more scientifically based negotiating positions. Such an effort will assist in the successful implementation of the FCCC (Framework Convention on Climate Change) and the Kyoto Protocol, for the negotiations in the post Kyoto process and in regulations to mitigate the possible consequences of climate change as concluded by IPCC. Very high resolution and better regional predictions are required to guide long term planning in sectors such as agriculture and energy. Results of simulations generated within the project will be available for other interested institutes in Europe, with the possibility of use in national projects on climate change impacts over the targeted areas.

Links with related projects (ENSEMBLES, CLAVIER, COST 734 – CLIVAGRI, QUANTIFY) were established through an active information exchange, some of these links are based on real cooperation and participation of some partners in these projects. The results of the climate change simulations generated within the CECILIA project will be available for other interested institutes, universities and research centres in Europe. Climate change impact data will be used for further impact research and policy studies and the results of these studies will be available as well. Both the reports and data generated during the project might contribute to the development of the next IPCC assessment report and the results will be shared and intercompared with other projects (ENSEMBLES, CLAVIER, COST 734, etc.).

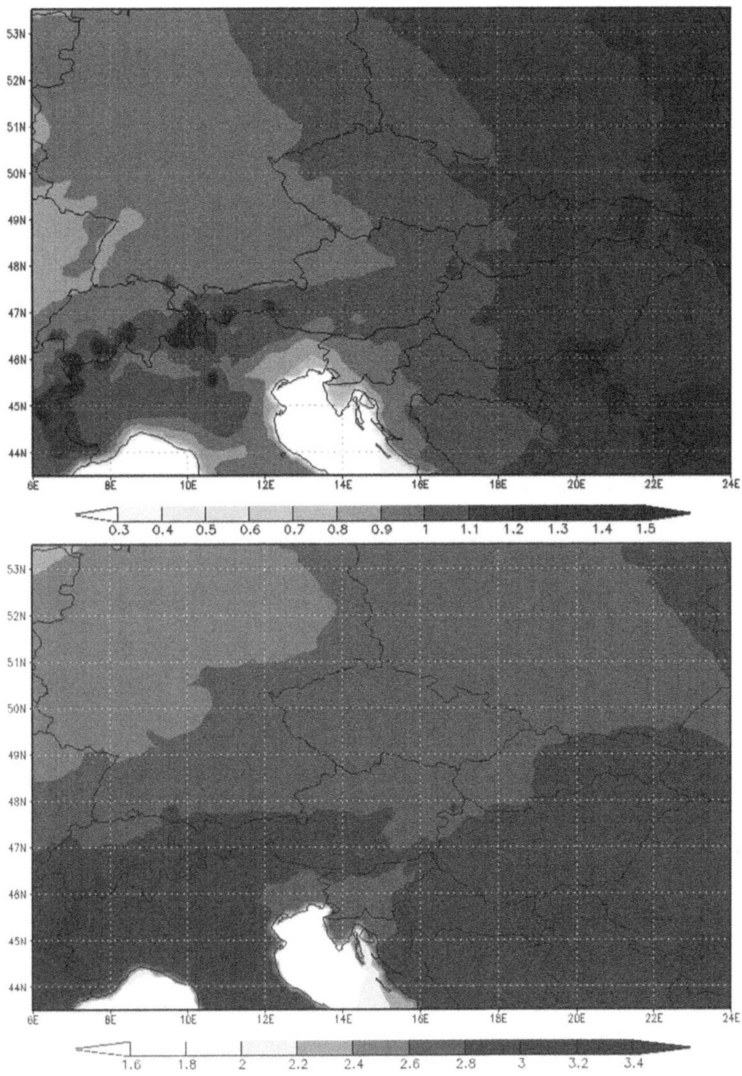

Fig. 11.7 Climate change signal for surface temperature in CUNI RegCM@10 km A1B simulations driven by RegCM@25 km runs for ENSEMBLES project (forced by ECHAM5 A1B scenario run) in terms of annual mean for mid-century time slice (2021–2050, *upper panel*) and end-of-century time slice (2071–2100, *bottom panel*)

There is one important point of the CECILIA dissemination and exploitation strategy, i.e. communication with end-users, stakeholders or policy makers to provide the necessary and reliable information directly as well as having the feedback to the project, especially from end-users for impact studies. By the end of the project, a final workshop will be held to present and disseminate its results. An important point of innovation consists in the fact that very high resolution climate information will allow application in integrated climate change impact studies, which will in turn provide for the first time necessary policy relevant information for decision makers and local authorities in the region. For more details on the progress in the CECILIA EC FP6 project see the project website http://www.cecilia-eu.org.

References

Beniston M, Stephenson DB (2004) Extreme climatic events and their evolution under changing climatic conditions. Glob Planet Change 44:1–9.

Elguindi N, Bi X, Giorgi F, Nagarajan B, Pal J, Solmon F, Rauscher S, Zakey A (2006) RegCM Version 3.1 User's Guide. PWCG Abdus Salam ICTP.

Frei C, Christensen JH, Deque M, Jacob D, Jones RG, Vidale PL (2003) Daily precipitation statistics in regional climate models: Evaluation and intercomparison for the European Alps. J Geophys Res 108(D3):4124.

Gates WL, Henderson-Sellers A, Boer GJ, Folland CK, Kitoh A, McAvaney BJ, Semazzi F, Smith N, Weaver AJ, Zeng Q-C (1996) Climate models – evaluation. In: Houghton JT et al. (ed) Climate Change 1995: The Science of Climate Change, Cambridge University Press, New York, pp. 229–284.

Giorgi F, Mearns LO (1999) Introduction to special section: Regional climate modeling revisited. J Geophys Res 104:6335–6352.

Giorgi F, Hewitson B, Christensen JH, Hulme M, von Storch H, Whetton P, Jones R, Mearns L, Fu C (2001) Regional climate information – evaluation and projections. In: Houghton JT, Ding Y, Griggs DJ, Noguer M, van der Linden PJ, Dai X, Maskell K, Johnson CA (eds) Climate Change 2001: The Scientific Basis, Cambridge University Press, Cambridge, pp. 583–638.

Gobron N, Pinty B, Melin F, Taberner M, Verstraete MM, Belward A, Lavergne T, Widlowski JL (2005) The state of vegetation in Europe following the 2003 drought. Int J Remote Sens 26:2013–2020.

Hafner S (2003) Trends in maize, rice and wheat yields for 188 nations over the past 40 years: A prevalence of linear growth. Agr Ecosyst Environ 97:275–283.

Halenka T, Kalvova J, Chladova Z, Demeterova A, Zemankova K, Belda M (2006) On the capability of RegCM to capture extremes in long term regional climate simulation – comparison with the observations for Czech Republic. Theor Appl Climatol 86(1–4):125–145.

Huntingford C, Jones RG, Prudhomme C, Lamb R, Gash JHC, Jones DA (2003) Regional climate-model predictions of extreme rainfall for a changing climate. Quart J Royal Met Soc 129: 1607–1621.

Huth R, Mladek R, Metelka L, Sedlak P, Huthova Z, Kliegrova S, Kysely J, Pokorna L, Janousek M, Halenka T (2003) On the integrability of limited-area numerical weather prediction model ALADIN over extended time periods. Studia Geoph Geod 47:863–873.

Kysely J (2004) Mortality and displacedmortality during heat waves in the Czech Republic. Int J Biometeorol 49:91–97.

Meehl GA, Tebaldi C, Nychka D (2004) Changes in frost days in simulations of 21st century climate. Clim Dyn 23:495–511.

Meehl GA, Tebaldi C (2004) More intense, more frequent and longer lasting heat waves in the 21st century. Science 305:994–997.

Menzel A, Jakobi G, Ahas R, Scheifinger H, Estrella N (2003) Variations of the climatological growing season (1951–2000) in Germany compared with other countries. Int J Climatol 23:793–812.

Menzel A, Sparks TH, Estrella N, Koch E, Aasa A, Ahas R, Alm-Kubler K, Bissolli P, Braslavska O, Briede A et al. (2006) European phonological response to climate change matches the warming pattern. Glob Change Biol 12:1969–1976.

Pal JS, Giorgi F, Bi X (2004) Consistency of recent European summer precipitation trends and extremes with future regional climate projections. Geoph Res Lett 31:L13202, doi:10.1029/2004GL019836.

Pal JS, Giorgi F, Bi X, Elguindi N, Solmon F, Gao X, Rauscher SA, Francisco R, Zakey A, Winter J et al. (2007) Regional climate modeling for the developing world: The ICTP RegCM3 and RegCNET. Bull Amer Meteor Soc 88:1395–1409.

Schär C, Vidale PL, Luthi D, Frei C, Haberli C, Liniger MA, Appenzeller C (2004) The role of increasing temperature variability in European summer heat waves. Nature 427:332–336.

Chapter 12
Vulnerabilities and Adaptation Options of European Agriculture

Josef Eitzinger, Sabina Thaler, Gerhard Kubu, Vesselin Alexandrov, Angel Utset, Dragutin T. Mihailovic, Branislava Lalic, Miroslav Trnka, Zdenek Zalud, Daniela Semeradova, Domenico Ventrella, Dimos P. Anastasiou, Mahmoud Medany, Samar Altaher, Janusz Olejnik, Jacek Leśny, Natalia Nemeshko, Michael Nikolaev, Catalin Simota, and George Cojocaru

Abstract Through a change in climatic conditions and variability, for example, certain extreme weather events (heat waves, droughts, etc.) are likely to occur more frequently in different spatial and time scales in future. Since agriculture is one the man' activities more dependant on weather behaviour, the impact on risks of agricultural production is indeed one of the most important issues in climate change assessments. Therefore an early recognition of risks and implementation of adaptation strategies is crucial as anticipatory; precautionary adaptation is more effective and less costly than forced, last minute, emergency adaptation or retrofitting. Results of climate change impact and adaptation studies often show considerable different results, depending on the spatial scale of regionalisation. However, for a decision maker, only a high spatial resolution of related study results is useful as it can represent local conditions and its spatial variability much better. This paper is based on the findings of the ADAGIO project (adagio-eu.org), which was focused on regional studies in order to uncover regional specific problems. In this context a bottom-up approach was used beside the top-down one of using scientific studies, involving regional experts and farmers in the evaluation of potential regional vulnerabilities and adaptation options. Results show, for example, that production risks, such as increasing drought and heat, are reported for most European regions. However, the vulnerabilities in the different regions are very much influenced by characteristics of the dominating agroecosystems and prevailing socio-economic conditions.

Keywords Climate change impacts on agriculture · Vulnerabilities in agriculture · Adaptation options

---

J. Eitzinger (✉)
University of Natural Resources and Applied Life Sciences, Vienna, Austria
e-mail: josef.eitzinger@boku.ac.at

V. Alexandrov et al. (eds.), *Global Environmental Change: Challenges to Science and Society in Southeastern Europe*, DOI 10.1007/978-90-481-8695-2_12,

12.1 Introduction

Global climate change will lead to shifts in climate behaviour and causes manifold impacts on ecosystems in the next decades. In particular, climate change will have significant effects on regional agricultural production (e.g., Alexandrov et al. 2002; Trnka et al. 2007, 2008), which has been considered as the most weather-dependent among all the human activities. Negative impacts on agricultural production could be avoided or reduced significantly by applying appropriate adaptation measures (e.g. in farm technology) supported by available impact models, as well as forecasts and warning systems for decision-making. This will secure sustainable agricultural production in the future as well.

Through a change in climatic conditions and variability, for example, extreme weather events (heat waves, droughts, etc.) will occur more frequently in different spatial and time scales in future. Since agriculture is one the man activities more dependant on weather behaviour, the impact on risks of agricultural production is indeed one of the most important issues in climate change assessments. On the other hand potential adaptation measures can provide a big potential to reduce risk and potential losses. As agriculture is an important economic factor in many countries, changing risk or potentials for agricultural production due to climate change will have a number of economic and socio-economic consequences (e.g. CAgM 1992, 1993; WMO 1995, 1997), such as loss of income in agriculture and food industry, significant higher costs for water and production techniques (e.g. irrigation systems), losses in the agroindustry.

Therefore an early recognition of risks and implementation of adaptation strategies is crucial as anticipatory; precautionary adaptation is more effective and less costly than forced, last minute, emergency adaptation or retrofitting (EEA 2004; ANL 1994; Parry and Carter 1998; Eitzinger et al. 2007). Moreover climate change may be more rapid and pronounced than current estimate suggest and there is a risk for under-adaptation. Immediate benefits can be gained by better adaptation to climate variability and extremes. At the policy level immediate benefits can be gained from removing policies and practices that result in ineffective adaptation. Finally climate change brings opportunities as well as threats, where potential benefits can be realised or increased by adaptation. All the above mentioned factors are very important to recognise as early as possible to ensure sustainable agricultural production systems and a living rural environment, especially for decision makers in their regional environments, including farmers.

Compared to the manifold potential impacts of climate change on agroecosystems, potential adaptation measures are even more complex because of the high number of options available through the human factor. Results of climate change impact and adaptation studies often show considerable different results, depending on the spatial scale of regionalisation. However, for a decision maker, only a high spatial resolution of related study results is useful as it can represent local conditions and its spatial variability much better.

12.2 Main Vulnerabilities of European Agricultural Regions

Regional studies and gathered feedback from experts and farmers show in general that drought and heat is the main factor having impact on agricultural vulnerability not only in the Mediterranean region, but also in the Central and Eastern European ones, except the north-eastern part of Europe (Russian regions). Another important aspect is that the increasing risk of pest and disease damages due to fast shifts of their occurrence may play a more important role for agricultural vulnerability as expected. However, till now this field is only rarely investigated in Europe. An important fact is also that there are increasing regional differences in the crop production potential in Europe due to climate change and that positive or negative impacted agricultural systems can vary in a relatively small spatial scale, depending on the specific limiting environmental conditions such as climate or soil conditions (especially in complex terrain).

12.2.1 Central European Region

In Central Europe agricultural production regions in Austria, Poland and Czech Republic were analysed regarding local vulnerabilities. Austria is characterized by small farm structures and big regional differences within short distances in the type of the prevailing agricultural production systems (mainly permanent grassland and dairy production, arable crop production and orchard as well as wine farming). This is caused by topographical conditions (Alpine region) and its impacts on regional climate, especially on temperature and precipitation conditions. These facts are also the reason why regional vulnerabilities in agriculture vary in a wide range within Austria. In recent studies it was shown that the more humid and cool alpine regions dominated by permanent grassland would increase their production potential, while the warmer and drier regions, dominated by arable crop production would face more drought and heat stress, having mostly negative impacts on production potential of summer crops. In this aspect not only the interannual variability of crop yields are expected to increase but also the differences between soils with high and low soil water storage capacity (increasing regional crop yield variability). Grassland dominated regions, which currently are close to the climatic limit regarding precipitation and grassland water balance, will face significant increase of production risk due to increasing drought and heat frequency. An example is shown in Figs. 12.1 and 12.2 showing the change of grassland regions under "low" production risk (of grassland and related dairy production) under climate scenarios in Austria and Czech Republic. Additionally the terrain (mean slope) and soil conditions are shown in Figs. 12.1 and 12.2, as adaptation options are limited in regions where soils or topography do not allow to change to arable crop production.

The most serious threat for Polish agriculture under future climatic conditions is considered to be the shortage of water. Based on historical sources it may be stated that the frequency of droughts in Poland is increasing. The Ministry of Agriculture

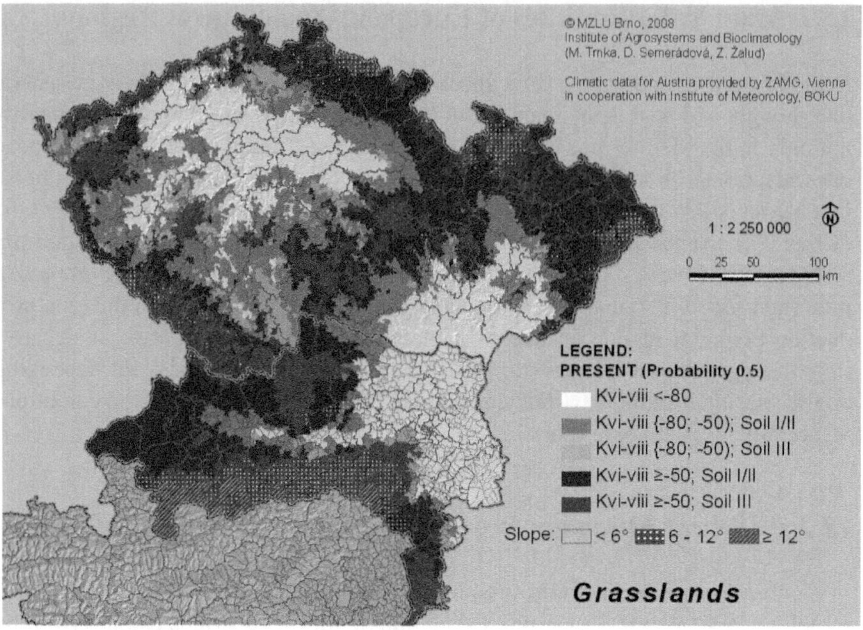

Fig. 12.1 Grassland dominated regions of different terrain and soil conditions (*grey scales*) with currently "low" production risk in Austria and Czech Republic, based on critical limits of mean grassland water balance (–80 and –50 mm) from June to August

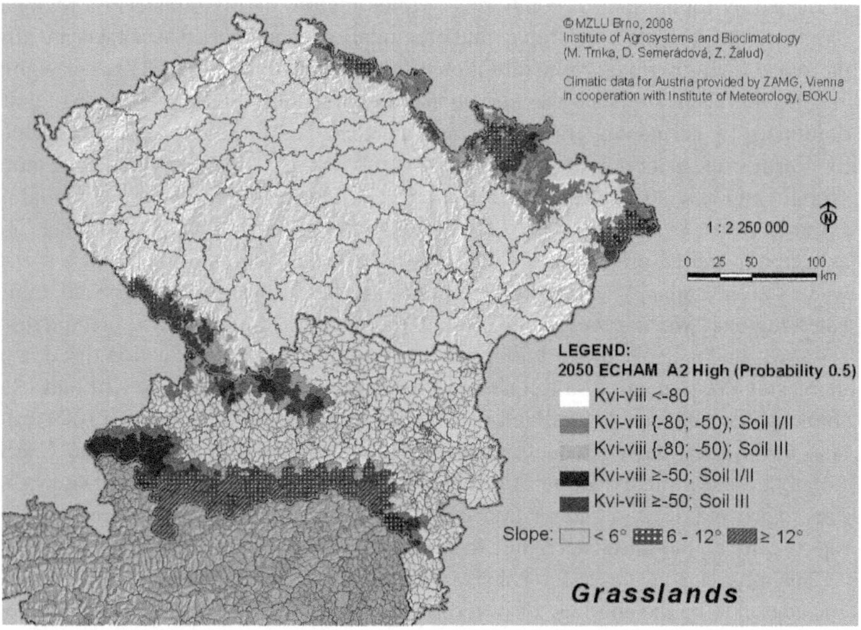

Fig. 12.2 Grassland dominated regions with "low" production risk in Austria and Czech Republic under the 2050's ECHAM A2 (extreme) scenario, based on the same conditions as in Fig. 12.1

and Rural Development assessed that as a result of drought in 2006 the biggest losses were found in meadows and pastures, mainly the 2nd and 3rd cuts – 40 to 100% losses, spring cereals – 20 to 60 % losses, winter cereals – 15 to 50% losses, rape – 15 to 45% losses, potatoes and sugar beets – 20 to 60% losses, vegetables – 30 to 60% losses. According to most climate scenarios, this type of extreme phenomena will occur more and more often and may be increasingly intensive. Thus situations similar to that in 2006 will have an increasingly adverse effect on agriculture. The forecasted warming will promote also a more intensive development of insects, which will cause their increasing distribution and population size, while for the same reason a more extensive development may be expected for fungal diseases and other plant diseases, especially in south-western Poland. A considerable weakness of Polish and Czech agriculture, not connected directly with climatic factors, results from traditional production methods maintained since the former socio-economic conditions. Even after the transformation in the 1990s, apart from the elimination of state ownership, the situation has not changed much. In the driest (south-east) regions of Czech Republic, for example, there is a lack of irrigation infrastructure and introduction of drought prone crops is relatively slow, so the farm economy often suffers from excessive dryness of the region. Especially polish agriculture is still dispersed and traditional to a considerable degree and requires in-depth structural transformations. Under such conditions the implementation of countermeasures for climatic changes may turn out to be problematic. Thus it is necessary for the agricultural policy of Poland in these countries to include the effect of climatic changes on agriculture.

12.2.2 Mediterranean Region

In the Mediterranean region selected areas of Spain, Italy, Greece and Egypt were investigated. Climate Change effects in Spain will posses e.g. new constraints on Castilla and Leon agriculture (selected region in Spain), as have been estimated from several assessments. Increment in the irrigation requirements and less available water could be expected. Besides, cereal cropping might be damaged by heat and water stresses in the spring, as well as vernalisation changes due to temperature increments in the winter. Moreover, high temperatures could significantly reduce wine quality. The farmers were asked by questionnaires about the Climate Change effects on their farms. About 39% of the farmers think there will be "Strong" effects, same percent consider there will be "Considerable" effects and 22% of them expect "Some" effects. None of the consulted farmers think that Climate Change will bring "no effect" to Castilla and Leon agriculture. The increment of pests and diseases is the worse climate risk identified by the regional farmers in this area.

In Italy the rainfall development under climate scenarios is uncertain. Considering the average all over Italy, there is a decrease of 5% per century in the annual precipitation amount, mainly in the spring season (−9% per century). The current prediction of General Climate Models downscaled with statistical or dynamical methods for the end of this century expect a reduction of summer precipitations in Southern Italy and little changes in Northern Italy with a small increase for the winter rains. Concerning the temperature, the scenarios show an increase quite uniform among the regions up to 5–6°C in Southern Italy in the summer months. Particularly large yield decreases are therefore expected for spring-sown crops (maize, sunflower and soybeans). For autumn-sown crops the impact is more geographically variable. Tubiello et al. (2000) found that, taking in account the positive effect of increased CO_2 and without adaptation of management and genotype, in Emilia Romagna the wheat and maize yield could decrease by 5–15%, soybean and barley yields by more than 20% and sorghum yields by more than 50%. In Puglia wheat yields are expected to decrease by 30–50%, sorghum yields by 10–30%. No productive changes were reported for sunflower.

The temperature increase could expand the suitable area for plants requiring high temperature such as grapewine, citrus and olive. The same can be expected for the cultivation of durum wheat and for most vegetables such as lettuce, fennel and cabbage for winter as well as spring cultivation, and melon. For the perennial crops, however, in the current production areas the yield variability (fruit production and quality) is expected to increase under global change scenarios. Such an increase in yield variability could have negative effects on the quality of products (fruit wine etc.) and, at the same time, could determine a higher economic risk for growers.

Olive is a typical Mediterranean crop particularly sensitive to low temperature and water shortage, thus both the northern and southern limits of cultivation are conditioned by the climate. Early flowering, determined by higher temperatures in winter period, can expose the plants to late frost risk if low temperatures occur. In the last decade in Northern part of Italy significant events of freezing took place with negative consequences in fruit and cereal sectors. However, the risk of damage to fruit trees caused by early autumn frosts is likely to decrease.

Actual trends in Greece show that olive, vine and winter cereals have higher adaptive capacity to lower precipitation and dry periods than cotton, maize, rice, tobacco and other industrial plants, which currently show an increase in agricultural area. Also many crops, especially maize and cotton, will need additional irrigation water resources under warmer and drier conditions. These characteristics could also mean a further increase of these crops in the coming decades, since especially their low requirements on time resources, while providing profit, could make them a more popular choice. For example, high quality wines and olive products are produced in Greek islands under very dry conditions, sometimes on marginal soils, and two of their most common diseases are associated with humidity, where drier conditions could lead to less frequent occurrence of them. Structural change in agriculture is known since the 1950s. If the farm profitability will decrease (such as due to climate change), areas with smaller farm size and no alternative income sources than agriculture may face population movements. Flatland agricultural areas close to major roads and urban centres, where agricultural farm size is 5 ha or generally larger than farm sizes in mountain regions, show a trend, which could be applicable for the future: part time farmers have more suitable conditions, since a shift to extensive production (e.g. from cotton to gardening products) is better suitable for a part time income. Too small farm size is a major cause of vulnerability, which influences profitability and it is an issue of governmental and organizational studies as well as actions. High profit crops, such as tobacco and cotton, show a decrease or huge reduction lately: the economic dependency of whole farming communities to these crops, creates the need for alternative crops with similar net profit levels, or the creation of local alternative sources of income.

12.2.3 Eastern Europe

In Eastern Europe, regions of Bulgaria, Russia, Serbia and Romania were investigated during the ADAGIO project. Agriculture is one of the most important sectors of the Bulgarian economy. Much of the Bulgarian population is occupied in it; however, the sector forms a relatively small share of the GDP. Bulgaria has excellent natural conditions for the development of agriculture. Cultivated agricultural land covers 48% of the total territory of the country. The favorable climate for various cultures, the fertile soils and long standing traditions in the sector, low labor cost, the presence of colleges and high schools on modern farming training can support a promising development of this sector. However, agriculture is in a crisis at present. Most of the farms are small and do not have at their disposal significant financial means. Various European funds are not enough efficiently used. The state must intervene to get out quickly of the crisis in this important structural sector of the Bulgarian economy. A survey shows that due to climate change in Bulgaria in the twenty-first century, main vulnerabilities of the agricultural sector will be expected for: (a) spring agricultural crops, due to the expected precipitation deficit during the warm half-year; (b) crops cultivated on infertile soils; (c) crops on non-irrigated areas; (d) arable lands in south-east Bulgaria, where even during the present climate, precipitation quantities are insufficient for normal growth, vegetation and productivity of agricultural crops.

Most vulnerable areas have been described, based on assessment of spatial and temporal trends in air temperature, precipitation, groundwater table, soil moisture evaporation, duration of warm season, dates of snow cover melting, weather extremes and agricultural production in last 25 years over European part of Russia. The main vulnerabilities of selected regions for Russia's agriculture under climate change were identified as follows:

North-West Russia:

- Increasing vulnerability of crops to insect pests and diseases including new occurrences for NW Russia;
- Availability of new, more southern weeds in crops;
- Increasing risk of snow mold on winter crops;
- Potential increase in risk of rotting out and water logging to winter crops due to increasing winter precipitation and warming;
- Dry spell hazards in early summer leading to reduced fertilizer efficiencies and need for additional irrigation.

Central Russia (and the same for Middle Volga area):

- Increase of ice crust and frost killing
- Impact of dry winds
- Spring and summer droughts
- Heat waves
- Water and wind erosion
- Floods

Lower Volga Region:

- Increase of frequency and duration of droughts
- Impact of dry winds
- Heat waves
- Water and wind erosion

The northern part of Serbia, where the province of Vojvodina was investigated, has one of the best natural agricultural production potentials in Europe. Main limiting factors that influence yield and quality of agricultural crop production are: low winter temperatures, late spring frosts, hail, temperature fluctuations during winter period, high summer temperatures and drought. Late spring frosts in Serbia cause significant damages in the majority of fruit species. Annual yield fluctuations occur as a result of frost damages of flowers, young fruits, especially in peach, apricot and plumbs. The whole territory of Serbia is often struck by thunderstorms with hail. Hail protection exists but does not always work, so damages occur. Droughts occur frequently in summer months (July, August). Lately, droughts have been occurred during the October–February period. Lack of winter precipitation threatens the production of winter cereals, maize, sugar beet, rapeseed. Wheat and barley yields are reduced by drought as well as increased temperatures in June. High temperatures in July (over 35°C) reduce pollination in maize and soybean. Increased temperatures in the month of March speed up vegetation so late frosts damage the fruits, sugar beet, etc. Lack of winter precipitation combined with summer droughts decreases the production of all crop species.

The most vulnerable agroecosystems in Romania are crop farming dominated by cereals (winter wheat – Fig. 12.3 – and maize) and crop farming for oilseed crops in South of Romania. The vulnerabilities detected in the dominating agroecosystems are:

- The small areas of agriculture exploitations (mainly in counties with low average income)
- Reduced number of crops (winter wheat, maize, sunflower)
- Decreasing level of water availability that is not counter balanced by introducing management practices for water conservation in dry farming (winter water harvesting, minimum tillage, using new cultivars fitting better with the climate changed conditions)
- Conservative attitude of rural population

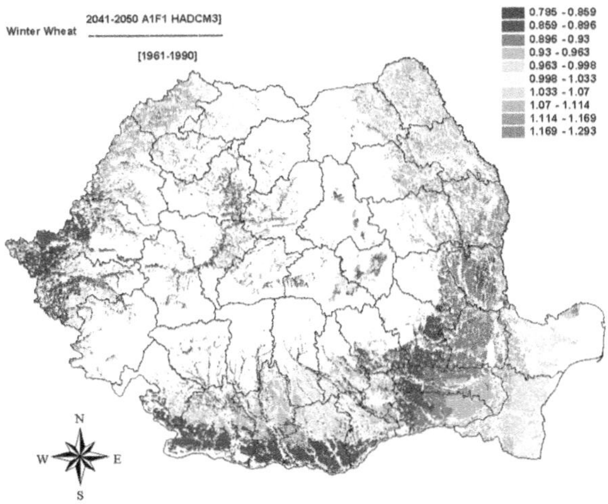

Fig. 12.3 High resolution map on the relative change of winter wheat production potential in Romania 1961–1990 vs. 2041–2050 (most *dark areas* represent regions with reduced yield potential)

Probably on a short and medium time horizon (2020) changes in rural space induced by introducing CAP and aggregation of land will be more important than the changes induced by climate changes.

12.3 Main Adaptation Options of Vulnerable European Agricultural Regions

In the European Union (27 countries) the number of agricultural holdings totals about 14.5 million. These farms manage 172 million hectares, or 61% are arable land. About 69% of the holdings in the EU (27) cultivate less than 5 ha. As already mentioned there is a wide range of variation of agricultural structural parameters between the EU countries, which have to be considered in the assessment of regional vulnerability and adaptive capacity. For example, there is a large range of mean farm size, land use and farm types. Main differences can be observed between western and eastern (former soviet union) countries in the number of different farm types per country and only partly by mean farm size. These structural variations overlap with climatic regions of Europe, which combination strongly influence production conditions.

12.3.1 Central European Region

The Central European region covers a wide range of climates ranging from cool and humid (e.g. alpine region) to warm and semiarid/humid (mostly lowland areas) of the transition of continental type to Atlantic and to Mediterranean type of climate. The ADAGIO countries Poland, Czech Republic and Austria can be considered as representative for that region. This region also covers a wide and distinctive difference in farm structure and socio-economic conditions, mostly caused by the different historical socio-economic as well as political developments of the western (Austria) and eastern (Poland, Czech Republic) countries. Potential adaptation options of the relevant agroecosystems are affected by both conditions.

In all countries of Central Europe all main arable crop production regions are affected by increasing drought conditions and water shortage during the summer period, which leads to several recommended adaptation options regarding the protection and efficient use of the agricultural water resources.

Agricultural grasslands (in combination with dairy farming) in regions below approximately 800 mm annual precipitation were shown as most vulnerable to the warming trend in Central Europe (see above), which comprise relatively large regions in Central Europe. Because of the fact that grassland production systems are much less flexible than arable farming systems regarding to a change in production techniques it is considered as the most critical sector and emphasis has to be paid on developing feasible adaptation options. For example, for regions, where a change to crop production or other alternatives by changing land use is difficult due to terrain or soil conditions, there are only few potential adaptation options possible for farming alternatives.

Perennial cropping systems, such as orchards and vineyards, will be probably mostly affected by changed pest and disease occurrence in Central Europe, which leads to the conclusion that adaptation measures should focus on this topic. Also adaptation options to avoid soil erosion and to improve crop management are a critical frequently reported issue for perennial crops.

Adaptation options mostly recommended in the Central European region are:

In the short term:

- Soil water conservation techniques (mulching, reduced and minimum tillage)
- Improving irrigation scheduling (e.g. by using agrometeorological stations)
- Change to heat tolerant cultivars
- Change cultivars and crops according temperature demands
- Change sowing date and shift timing of field works
- Improving and introducing monitoring systems for pest and diseases
- Adaptation in crop rotation (e.g. more winter crops)
- Ensure frost protection methods (and for hail in some regions)
- Effective insurance system (ideally supported by government)
- Adapting animal stables for heat waves, ensuring power generation and increasing hygienic measures for farm food production (e.g. milk production)

In the midterm:

- Improving or establishing irrigation infrastructure
- Breeding of adapted cultivars (e.g. for higher water use efficiency and heat stress)
- Increasing crop diversification (farm and regional scale)
- Increasing storage capacities (fodder storage)

- Change of land use and/or production system (e.g. cereals to maize, grassland to vineyards, grassland or crops to energy biomass production)
- Landscape structure improvements for reducing evapotranspiration (by lowering the wind speed, in some flat semi-arid regions)
- To ensure higher market price stability for agricultural products (which significantly can improve sustainable planning and implementation of adaptation options) by using appropriate tools (e.g. political measures, ecological farming, establishing regional food production and local markets, develop marketing concepts, such as "terroir" characteristics for wine quality aspects).

Mostly named limitations of adaptation options in Central Europe, related to socio-economic conditions are structural changes, which are mainly driven by economic reasons rather than by climatic ones. Among these are a strong decrease of income per work unit (esp. Poland), too small farm sizes (Austria, Poland) or complex terrain limitations and high production costs (Austria). Too low income at small farms (esp. related to family business) leads to abandoning small farms (Austria, Poland) or reduction of agricultural used land (Poland) in all regions. Farmers of small farms are often also less educated in agricultural production (often part time depending on other incomes) leading to a decreasing success or interest in farming.

The adaptive capacity of small farms seems to be better in well developed countries with better infrastructure or well established local markets such as in Austria, where already many small farms could successfully change to ecological production or alternatives and niches. Eastern countries (see also Section 12.3.2) are much less flexible in that sense, probably also by the missing political support, low financial resources, not well developed infrastructures, fewer possibilities of additional incomes beside agriculture at the same location.

Main other limitations in Central Europe are related to bad existing or destroyed infrastructure (esp. irrigation systems) and less financial resources for improvements in the former socialistic countries. Often too less water is available in the region for additional irrigation (partly an infrastructure problem), also in the "western" countries of Central Europe. Other limitations are too high costs of certain adaptation options, terrain and soil restrictions, no (or still no) market for the specific product. Market price uncertainties lead to increasing risks for sustainable implementation of adaptation measures; especially those which are related to investments (e.g. change of land use or production system). In several cases labour and time pressures are also limiting the implementation of adaptation options (such as a change to ecological farming in Austria).

12.3.2 Eastern European Region

Bulgaria, Serbia, Romania and Russia, representing Eastern Europe, are characterized by various climatic conditions from continental climate in the European part of Russia to mild climate in South Bulgaria, which is very near to the Mediterranean region. Agriculture in these countries is still in the transition of the market economy as a result of social and political changes during the last 20 years. The potential adaptation measures of the respective agroecosystems are influenced by the specific climatic conditions (see Section 12.2.3) and strongly limited by the prevailing socio-economic conditions. However, there are significant differences in the agricultural structure and the conditions of the agricultural infrastructure between these countries. In Bulgaria and Romania a main problem is the decrease of rural population (which is an effect of EU-membership of the past years), leading to a lack of man power, land abandonment and finally to a further weakening of agricultural infrastructure. In Bulgaria and Russia especially the (past) large farm sizes infrastructure, still not adapted to the new market system, is an additional problem. Serbia seem to be still have the best working agricultural system in these countries; however, in several sectors modernization and financial support is strongly needed to improve future competition strengths in the agricultural sector.

As an example, the situation of irrigation infrastructure in Bulgaria can be described as follows:

Actual problems and irrigation history in Bulgaria:

- Irrigation activities in Bulgaria have a long tradition dating back to the fifteenth century when first rice production along the Maritza River and, later, vegetable and fruit growing were supported by means of irrigation.
- Irrigation is very important for Bulgarian agriculture, but until the end of the Second World War only a small part of the land was irrigated.
- During the1960s, the state initiated an extensive program to increase irrigated areas. Since cooperatives dominated the organisational form, irrigation systems were designed to supply water to large production units.
- The main sources of water supply were large dams, located in the mountains, and rivers. Groundwater was used as a complementary source.
- In 1990, the total irrigated land was about 1,200,000 ha (25% of the arable land in Bulgaria).
- The land restitution process was slow and contradictory.
- At the end of land reform in the year 2000, Bulgarian farm structure was dominated by three groups: small subsistent farms operated by people close to retirement, cooperatives, most of them in bad financial situation, and large commercial farms.
- The number of middle-size family farms remained small.
- During transition, the amount of water used for irrigation in Bulgaria has sharply declined.
- In addition, the share of actually irrigated areas to those that can be irrigated is low.
- Large sections of existing irrigation systems lie abandoned, and the ones still in use are barely maintained.
- Crops, such as wheat and barley, have replaced more water intensive crops, including vegetables, rice and maize.
- Irrigation, until recently a major water user in Bulgaria, has been drastically affected.

- Uneven distribution of Bulgaria's natural water resources over time and space makes irrigation necessary to reduce production risk and insures that the common-pool resource retains continuous in high economic importance.
- Yet, the irrigation systems were built to serve large production units during socialism and do not meet the needs of the huge number of small-scale landowners that emerged following the land restitution process.
- Moreover, facilities have largely deteriorated; property rights on the infrastructure are ambiguous and water loss in the irrigation systems amounts to 70% owing to no maintained facilities and water stealing.

Beyond the socio-economic shortcomings, most adaptation options in Eastern European agriculture are directed to an improved water management of cropping systems (except the North-Western part of Russia). Improving irrigation systems and infrastructure, water harvesting methods, adapting crop rotation, changing cultivars or cropping pattern and others are often recommended to increase efficient water use, production as well as production stability in agriculture. An improved pest, disease and weed management is a common recommended adaptation measure for all regions in Eastern Europe.

Table 12.1 gives an overview reported examples of the various recommended adaptation options for different (climatic) regions in Eastern Europe.

12.3.3 Mediterranean Region

In general, one of the most important effect of climate warming is the increase of the evaporative demand of the atmosphere. Consequently, an increase of irrigation requirement is expected for the next decades of this century. However, this effect is counterbalanced by the shortening of phenological cycle length for most determinate plants. Therefore several studies indicated only relatively low increases of irrigation demand and in several cases also a reduction at the seasonal scale.

Advancing the sowing time reduces the risk to have higher irrigation requirements with low values of water use efficiency. Optimization of sowing/transplanting time (advancing for spring crops – delaying for winter crops) and the selecting varieties and/or species among those available at present are the most feasible and fast adaptation options to reduce the negative effects of climate change.

However, the main factor that characterizes the Mediterranean region is proposed adaptation strategies concerning irrigation and, in general, the water requirements of the cropping systems cultivated. The optimization of amount and time of irrigation and/or other water management practices is the main way to increase the water use efficiency at farm/regional level. The adaptation at farm/regional scale to new technologies to schedule the irrigation, of advanced procedures based on remote sensing information and structural enhancement for using urban depurated water are the mostly proposed strategies of adaptation. In same cases, such as for the Nile delta

Table 12.1 Examples of reported aspects of regional adaptation options in Eastern Europe

Domi-nating agroeco-system	Region	Observed trends in adaptations to climate change	Recommended feasible adaptation options to climate change	Identified limitations for adaptation options to climate change	Uncertainties, cost/benefits, risks
Grape/ Vineyards	South of European; part of Russia	Maintaining heat and water regime in optimum; control of surface runoff	More effective application of techniques for frost protection; slope terrassing		Changes in pest and disease risk
	Northern Bulgaria	Earlier sowing and harvesting	Dry resistant cultivars (>2020s); irrigation changes	Topographical limitations; farmer behaviour; legislation barriers	Change to other crops increases flexibility; higher costs for machinery; farmer incomes
Cereals	Northern Vojvodina	Decrease land under wheat and increase land under barley as well as triticale; increase of farmers interest for climatic change issues and enhanced farmers interest for expert advices; insufficient support by government; increasing involvement of experts and advisory services	Optimized soil cultivation and sowing timing; adapted plant density; judicious use of NPK fertilizers; maintenance of good plant health; incorporation of plant residues into the soil	Low farmers income; unfavourable bank credits; no subventions for plant production; high market price fluctuations; still ongoing privatization of food industry; still no national adaptation strategy; great number of farms is out of advisory service programs; old machinery	Suggested measures of adaptation do not increase production, however, application of these measures would increase production stability; farmers do not have influence on prices of input or output; association of farmers do not react adequately; crisis in animal production is also a related problem (food cereals)

Table 12.1 (continued)

Domi-nating agroeco-system	Region	Observed trends in adaptations to climate change	Recommended feasible adaptation options to climate change	Identified limitations for adaptation options to climate change	Uncertainties, cost/benefits, risks
	Romania (West – Banat)	Shift of timetable for tillage operations; several farms introducing minimum tillage; increase acreage for winter cereals (wheat, barley) and spring barley	Change timings of field operations; introducing minimum tillage; change of cultivar structure	–	Higher cost of machinery; high yield variability from year to year; training of farmers for using minimum tillage systems
	Southeast of European, part of Russia	Change autumn tillage; applying manures; improving crop rotation systems by introduction of alfalfa	Effective snow hedge application (water harvesting)	Solonetz-like soils; solonetz formation; sandy soils	Reduced frost killing risk

Table 12.1 (continued)

Domi-nating agroeco-system	Region	Observed trends in adaptations to climate change	Recommended feasible adaptation options to climate change	Identified limitations for adaptation options to climate change	Uncertainties, cost/benefits, risks
Maize	Southern Vojvodina	Greater percentage of early cultivars; adjustment of big farms to adaptation measures; enhanced farmers interest for expert advices	Use of plant rotation (example: maize sown after cereals); plant residues incorporation with nitrogen application; decrease of plant density on dry soils; judicious use of NPK fertilizers; optimized time and quality of seedbed preparation and sowing	Low farmers income; still high density of weeds on arable land (bad weed management); old machinery reduces soil cultivation quality; scarce information to the farmers	Crisis in animal production decreases local corn prices; high variation of incomes and prices increases economical risks; corn is more often devastated by summer drought or hail
	Chernozem zone (Central- Southern Russia)	Water conservation tillage; regional water management; afforestation; more productive cultivars	Slope terracing; introduction of new hybrids	Drainage systems limited in hilly terrain; water erosion; dry winds and droughts	Intensified harmfulness of corn bug
Orchards	Subotica	Bare soil system in orchards; moving orchards to less light soils; more anti- hail nets covers (shade benefits)	Integrated fruit production; enhancement of organic contents in soils; hail nets	High introducing costs; demand of manure	Less costs for plant protection; nutrition added value; benefits only in case of optimal application date; reduced production risks; machinery needed

Table 12.1 (continued)

Domi-nating agroeco-system	Region	Observed trends in adaptations to climate change	Recommended feasible adaptation options to climate change	Identified limitations for adaptation options to climate change	Uncertainties, cost/benefits, risks
Vegetables	Southern Vojvodina	Rapid construction and application of irrigation systems; Introduction of new plant cultivars; application of modern pesticides	Introduction of alternative species; more use of organic fertilizers; judicious use of NPK fertilizers (Nitrogen control); maintaining good health of plants; good water management	Low farmer incomes; undeveloped refining industry; chaotic market conditions; great number of small farms, insufficient number of special machinery; insufficient production of manure	Uncertain placement of agricultural products on market (cabbage, potato); unlimited import of vegetables; high farmers dependence on the trade sector
	North-West Russia	Improving resistances of potatoes cultivars to intensifying potatoes diseases	More productive cultivars	Forest-clad relief; bogginess; low soil fertility; soil acidification	Increasing distribution of insect pests, diseases and weeds; new insect pests

in Egypt, particular adaptation options are related to sea level rise, such as "switch of agricultural cropping to aquaculture" with positive impacts on the national food security, but inducing new environmental pressure on the natural resources on the region.

12.3.3.1 Adaptation Options in Short Term

Change of cultivars is an often named easy option of adaptation of climate change. The criteria taken into consideration would be based on more appropriate thermal time and vernalisation requirements; increased resistance to heat waves and increased resistance to drought. Cultivars with higher thermal time requirement and consequently with longer growing cycle, represent an adaptation measure in order to reduce the negative impact of climate change.

Change of crops has to be established with the same criteria of changing cultivar. This option was considered for grassland systems changing to annual fodder crops. In cropping systems based on cereals and industrial crops, where a reduction of the availability of irrigation water can be expected, an increasing cultivation of winter wheat or other winter cereals can be recommended instead of summer crops. In Northern Italy, for example, there is an increasing substitution of soft winter wheat by durum winter wheat and this tendency could increase in the future. In vegetable rotations, above all in Southern Italy, a shift toward winter vegetables (lettuce, fennel, cabbage) is expected as well.

The optimization of water use in agricultural production is the most important factor to reduce the negative impact of climate warming on local water resources, i.e. improving irrigation scheduling. To optimize the amount and time of irrigation means to reduce the water losses by deep percolation and soil evaporation. Giving water at the right moment avoids water stress for the plants and drainage losses of water and nutrients. The recommended irrigation scheduling methods are the evapotranspirometric method, monitoring of the water status of soil and plant with tensiometers, FDR and TDR technology, deficit irrigation and partial root drying.

Localized irrigation methods, such as drip or microsplinker irrigation, are very common in the Mediterranean countries for tree crops (vineyards, citrus and partially olive) and also vegetables (tomato, water melon, lettuce, cabbage, etc.). In Northern Italy the prevailing methods are still sprinkler, flood or furrows irrigation. New recommended methods are for example low energy precision irrigation (LEPA), Ultra Low Drip Irrigation (ULDI) with 0.1–0.2 $l\ h^{-1}$, irrigation with photovoltaic energy and sub-irrigation.

The potential productivity of more humid zones in the Mediterranean region is in general expected to increase due to longer vegetation periods. If advanced sowing time is applied, it is appropriate to recommend changes in fertilisation schemes in annual crops and perennial systems, such as grassland and trees, and in particular vineyards. Fertirrigation is also an important opportunity.

Tillage practices and soil cultivation methods (e.g. mulching systems or reduced soil cultivation, minimum tillage) have important effects on soil water conservation and can protect the soil against soil erosion by water and wind. The introduction of relevant methods are expected to be useful especially for cropping systems based on cereals with particular reference for winter wheat and maize, as well as in hilly terrain.

Other options to reduce soil water losses include soil mulching with white synthetic or, preferably, vegetal material as crop residues from the previous cultivation.

Where there is limited availability of water of good quality for irrigation, saline waters represent an important resource. This is typical in several areas of Southern Italy with particular reference to coastal areas of Adriatic and Ionia Sea. In general this water comes from not very deep groundwater of unconfined or perched types and represents the only available resource of water in these areas. In other cases, the farmers generally use the water from irrigation networks, often coming from rivers (such as in Spain or Egypt) and water from wells when the first option is not available.

There are recommended significant changes in crop protection methods (pest, diseases and weeds) for any crops. The economic benefit of monitoring and protection is different in function of many factors. Especially for arable cropping systems it is an important option and it becomes strategically very fundamental for tree and vegetables crops like vineyards, orchards and vegetables such as tomato.

12.3.3.2 Adaptation Options in Midterm

Breeding of new varieties and increasing cultivar diversification is a most important task concerning appropriate thermal time and vernalisation requirements, increased resistance to heat waves and drought, increased resistance to new or expected dangerous pests, diseases and weeds and in general improved productivity or yield stability.

The structural enhancement of irrigation efficiency is another important factor for countries such as Italy, Greece and Spain. In general in many areas improvements of canalisation systems are required in order to reduce the relevant water losses which can add up to 40–50% in some places.

Changes in farming systems such as the noted and expected tendency of partly substitution of food production by energy production, considering that several crops are suitable for different type of biofuels and a northward expansion of high productive crops.

For tree crops the temperature increase will expand the suitable growing areas for plants requiring high temperatures such as grapewine, citrus and olive. The same can be recommended for the cultivation of winter durum wheat in substitution of soft winter wheat in some locations such as in Northern Italy. Also a shift to vegetables in winter and spring cultivation such as lettuce, fennel, cabbage and melon can be recommended for the warmer future climate.

Significant increases of water use efficiency can be obtained by introducing of new technologies to schedule irrigation. Remote sensing information should be applied in order to improve the land cover classification (very important for water resources planning), the assimilation of leaf area index (LAI) and shallow soil water contents in order to support estimation of regional soil/crop water balance and to schedule the irrigation at watershed scale with daily temporal resolution.

Finally, structural enhancements at regional scale should be directed to use depurated urban water. This is an important question that needs additionally legislative support because national (or European) law is very restrictive for depurating this type of waters with very low contamination thresholds values requiring high costs for cleaning these waters (Table 12.2).

Table 12.2 Overview of the most feasible adaptation options in the Mediterranean region

Adaptation options that are feasible in short term	Adaptation options that are feasible in midterm
Optimization of sowing/transplanting time (advancing for spring crops – delaying for winter crops); cropping activities	Constitution (with traditional or innovative breeding methods), evaluation and introduction of new varieties
Selecting varieties and/or species among those available at present with more appropriate thermal time and vernalisation requirements and/or with increased resistance to heat waves and drought	Structural enhancement of irrigation efficiency and changes in farming systems Adoption at farm/regional scale of new technologies to improve irrigation scheduling for obtaining high water use efficiency
Optimization of amount and time of irrigation and/or other water management practices such as drainage	Adoption of advanced procedures based on remote sensing information in order to optimize the use of water resources at regional scale
Optimization of fertiliser rates to maintain grain or fruit quality	
Applying available technologies to conserve soil water (no tillage, minimum tillage, reduced soil cultivation, mulching systems)	Structural enhancements at regional/farm scale for using urban depurated water
Using water resources of low quality (saline waters)	
Improve the effectiveness of pest, disease and weed management practices	
Particular adaptation strategies for Egypt (Nile delta)	
Improvements of the current irrigation and drainage system	Switch cropping activates to aquaculture
	Environmental controlled production techniques

12.4 Conclusions

The results of the ADAGIO project show a complex picture of the vulnerabilities and potential adaptation options to climate change in agriculture in different European regions. Although dominating risks, such as increasing drought and heat, are similar in all regions, the vulnerabilities in the different regions are very much influenced by characteristics of the dominating agroecosystems and prevailing socio-economic conditions. This is even more significant for potential adaptation measures at the different levels, which have to reflect the regional conditions.

Further needs to improve adaptation options for the future in most venerable regions of Europe concern especially an improvement and wider implementation of effective irrigation systems and methods (e.g. microsprinkler and drip irrigation; an increase of supplemental irrigation for normally rainfed crops) and improvement of irrigation infrastructure, including using of new water resources such as waste waters or repair leaking of pipes.

The most important challenge for irrigated agriculture will be to save water and to increase the water use efficiency in the whole irrigation chain. One of the reasons to increase the water use efficiency is linked to economical aspects because the competition with the other productive and/or social sector will determine an increase of water price in the future.

In rainfed agriculture water conservation techniques are increasingly important to reduce drought risks of crops in Europe. Measures such as reduced soil cultivation, no- or minimum-tillage, mulching systems or adapting crop rotation have further to be investigated in detail in the context of crop rotations, soil conditions and climatic regions to develop optimized recommendations for local applications.

Much emphasis has to be paid to develop recommendations for adaptation measures for permanent grassland systems and dairy production, as low flexible systems will be most affected by climate warming and dryer conditions. In general, but especially for perennial crops, further research is strongly needed for improved pest, disease and weed management, i.e. to develop infrastructures for an early recognition of risks (e.g. monitoring systems) and proactive measures of spatially shifting occurrences according to the shifts in agroecological zones which will appear during the next decades.

In the midterm crop breeding for more stress tolerant crops and a wider crop and cultivar diversification has to be forced in order to reduce yield risks. Beside these measures, conditions for less variable market prices of agricultural products should be established, based on concepts such as regional food production or ecological farming in order to ensure sustainable agriculture and successful implementation of adaptation measures to climate change.

Acknowledgments This study is supported by the EU-Project "Adaptation of agriculture in European regions at environmental risk under climate change" (ADAGIO – www.adagio-eu.org).

References

Alexandrov V, Eitzinger J, Cajic V, Oberforster M (2002) Potential impact of climate change on selected agricultural crops in north-eastern Austria. Glob Change Biol 8(4): 372–389.

Argonne National Laboratory (ANL) (1994) Guidance for Vulnerability and Adaptation Assessments. U.S.C.S.P. USA, 205 pp.

Commission of Agricultural Meteorology (CAgM) (1992) Monitoring Assessment and Combat of Drought and Desertification. WCP/CAgM, Report No. 47, WMO/TD 505, Geneva 112 pp.
Commission of Agricultural Meteorology (CAgM) (1993) Report of the RA II Working Group on Agricultural Meteorology, Part IV: Drought and Desertification. Report No. 52, WMO/TD-No. 524, pp. 1–43.
Eitzinger J, Utset A, Trnka M, Zalud Z, Nikolaev M, Uskov I (2007) Weather and climate and optimization of farm technologies at different input levels. In: Sivakumar MVK, Motha R. (eds) Managing Weather and Climate Risks in Agriculture, Springer, Berlin.
European Environmental Agency (EEA) (2004) Impacts of Europe's changing climate: An indicator based assessment. EEA Report No 2/2004.
Parry M, Carter T (1998) Climate Impact and Adaptation Assessment. Earthscan Publication Ltd, London.
Trnka M, Eitzinger J, Semeradova D, Hlavinka P, Balek J, Dubrovsky M, Kubu G, Stepanek P, Thaler S, Zalud Z (2008) Expected changes of agroclimatic conditions in Central Europe. Clim Change (submitted).
Trnka M, Muska F, Semeradova D, Dubrovsky M, Kocmankova E, Zalud Z (2007) European Corn Borer life stage model: Regional estimates of pest development and spatial distribution under present and future climate. J Ecol Model (207): 61–84.
Tubiello FN, Donatelli M, Rosenzweig C, Stockle CO (2000) Effects of climate change and elevated CO_2 on cropping systems: Model predictions at two Italian locations. Euro J Agron 12:179–189.
World Meteorological Organization (WMO) (1995) The Global Climate System Review. No. 819, 150 pp.
World Meteorological Organization (WMO) (1997) Climate, Drought and Desertification. Geneva, WMO No. 869.

Chapter 13
Projection of Climate Change for South East Europe and Related Impacts

Eglantina Bruci

Abstract This paper reviews climate change projections based on a range of emissions scenarios extending up to the end of the twenty-first century. The study area covers South East European countries, such as Croatia, Bosnia and Herzegovina (B&H), Serbia, Montenegro, Albania and the Former Yugoslavian Republic of Macedonia (FYROM). Low- and high-resolution projections of future changes in temperature and precipitation are analyzed. Scenarios exhibit an expected increase in annual air temperature of more than 5°C with very slight differences over the region. The annual temperature is likely to increase by 5.3°C over Serbia and northern Croatia, 5.4°C over Bosnia and Herzegovina (B&H), rising to an increase of 5.5°C through inland parts of Montenegro, as well as northern and central parts of Albania, and 5.3°C in the FYROM. Coastal areas are likely to be less warm (a very slight change indeed) than those in the rest of study because of the equalizing effect of the sea. Less total precipitation, with an irregular pattern of change, is likely throughout the study area for the 2080s. A reduction of about 16–20% is expected in the northern part of Croatia and 13–16% along the coast, rising to about 22% through Montenegro and Serbia. The expected reduction will intensify over Albania (around 25% in the north, 21% in the central region, 24% in the south, and 20% in the coastal area) and Macedonia. As a result of the reduction in annual total precipitation, the study area could experience a general decrease in runoff and demands for water could increase, especially in summer. It could cause serious problems in hydropower energy production in countries like Albania which depend heavily on hydropower.

Keywords Climate changes · Climate projection · GCM · Scenarios

E. Bruci (✉)
Climate Change Programme, Ministry of Environment, Forests and Water Administration, Tirana, Albania
e-mail: eglantina.bruci@undp.org

V. Alexandrov et al. (eds.), *Global Environmental Change: Challenges to Science and Society in Southeastern Europe*, DOI 10.1007/978-90-481-8695-2_13,

13.1 Introduction

Warming of the climate system is unequivocal, as has become evident from observed increases in global average air and ocean temperatures, widespread melting of snow and ice, and the rising global average sea level (IPCC 2007). The world's average surface temperature has increased by around 0.74°C over the past 100 years (1906–2005). A warming of about 0.2°C is projected for each of the next two decades (IPCC 2007).

What are the likely projections for the future? The study area in this paper covers South East European countries, such as Croatia, Bosnia and Herzegovina (B&H), Serbia, Montenegro, Albania and the Former Yugoslavian Republic of Macedonia (FYROM).

13.2 Method

Low- and high-resolution projections of future climate have been analyzed in order to understand the level of detail and variation among them. Analysis consists of assessing:

- likely annual changes in temperature and precipitation for the 2080s (years 2070–2100) running MAGICC for different IPCC scenarios, with HAD300 as the global circulation model (GCM).
- likely annual changes in temperature and precipitation for the 2080s, running MAGICC for the same scenario A2ASF under different GCMs.

High-resolution regional data (with horizontal resolutions of 50 km) for actual climate and climate change projections are provided by the Hadley Centre.[1]

The expected changes are calculated as deviations from the mean value (with a 1961–1990 referral baseline) for temperatures, and in percentage for the other elements.

Analysis continues with the assessment of expected annual and seasonal changes for the 2080s for different climatic elements such as air temperature at 1.5 m, total precipitation, specific humidity, relative humidity, potential evaporation, soil moisture content, and wind speed. The trend of these changes is evaluated over the study area. This paper reviews analysis for only two factors; air temperature at 1.5 m and precipitation.

---

[1](The Hadley Centre's version of the RCM (HadRM3P) is based on HadAM3H, an improved version of the atmospheric component of the latest Hadley Centre coupled AOGCM, HadCM3, run under the SRES A2 emission scenario).

13.3 Expected Climate Changes

A climate change projection is the difference between a model simulation of the present climate[2] and the model climate projection for a period (in this note generally taken to be the period 2071–2100) in the future, under a specific emissions scenario.

13.3.1 Low-Resolution Projections of Future Climate

13.3.1.1 Use of Different Scenarios

To analyze expected changes in temperature and precipitation for the 2080s (years 2071–2100), MAGICC is run for different IPCC scenarios (AIM, A2ASF, B1 IMA, B2MES, A1FI, A2AIM, A2A1MI, AVE, A1B), with HAD300 as the Global Circulation Model.

Patterns of temperature change are similar in all scenarios, with a warming everywhere over the study area which increases in A1FIMI, A1BAIM, and A2ASF. The least warming is projected by A2AIM, A2A1MI, B1MES, and B2MES (Fig. 13.1).

Quite the reverse pattern is seen when analyzing expected changes in the precipitation total. All scenarios used lead to a relatively high decrease (higher than −9%) in annual precipitation totals (Fig. 13.2) over the study area. Quite a different situation – with plenty of precipitation – is expected in northern Europe.

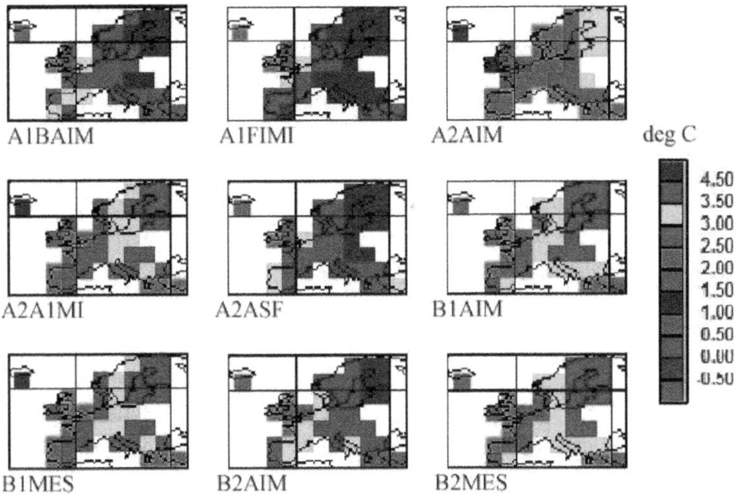

Fig. 13.1 Annual temperature changes (°C) for the 2080s, for different climate scenarios, region Europe

[2]In this report, the period 1961–1990 is considered to be the climatic baseline.

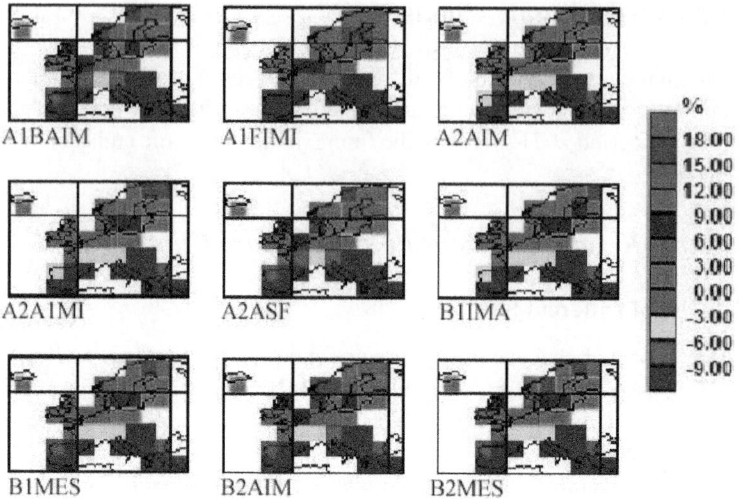

Fig. 13.2 Expected annual precipitation changes (%) for the 2080s, HAD300, different climate scenarios

Taking these results into consideration, it might generally be concluded that different scenarios project different changes, but in every case the tendencies are the same, a likely increase in temperature and decrease in precipitation.

Analysis shows that the highest increases in annual average temperature are projected by A2ASF, A1BAIM, B2AIM (respectively 4.3, 4.0, and 3.8°C), and the lowest by B2MES, and A2AIM (2.3 and 2.7°C). As for precipitation, scenarios A1FIMI, B1MES, and A2AIM project the highest decrease in annual precipitation. It is also better to make climate projections for a range of SRES emissions scenarios in order to reduce uncertainty.

13.3.1.2 Use of Different Global Circulation Models

An analogue situation is observed by analyzing outputs generated by running MAGICC/SCENGEN for A2ASF with different GCMs. Likely changes are presented in Fig. 13.3 (changes referred to in the gridbox with latitude 40.0N – 45.0N and longitude 20.0E – 25.0E). It is clearly shown that global climate models used, which contain different representations of the climate system, project different patterns and magnitudes of climate change for the same period in the future when using the same concentration scenarios. Thus, expected annual changes in temperature over the study area vary from +2.5°C to up to more than +4.5°C. As for annual precipitation, its total is expected to drop from −3% to more than −9%. The greatest warming scenario is projected by HAD300 and ECH498, and the largest precipitation decreases by ECH395 (about −29%) (Fig. 13.3).

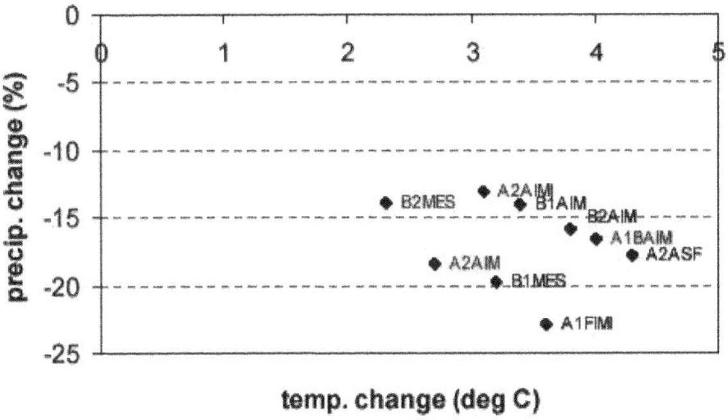

Fig. 13.3 Future changes in annual temperature and precipitation under different GCMs for the 2080s, gridbox: 40.0N – 45.0N; 20.0E – 25.0E

13.3.2 High-Resolution Projections of Future Climate

The main conclusion in Section 13.2.1 is that in general projected tendencies are for a likely increase in temperature and decrease in precipitation. It is to be stressed that because of the low resolution in every case, land appears to warm at the same rate as the sea. Even large islands such as Corsica, Sardinia and Sicily are not seen. The situation seems to be rather different when RCM (HadRM3P) is used.

The following analysis takes into consideration not only future temperature and precipitation projections, but also some other closely related elements.

13.3.2.1 Air Temperature (at 1.5 m)

Annual Changes

The study area is likely to become warmer for the 2080s (Fig. 13.4). The scenario/model exhibits an expected increase in annual air temperature of more than 5°C over the region. The differences from one gridbox to another are very slight. So we may notice that temperature is likely to increase by 5.3°C over Serbia and north Croatia, 5.4°C over Bosnia and Herzegovina (B&H), reaching a value of 5.5°C through inland parts of Montenegro, and north and central parts of Albania and 5.3°C in the FYROM. Coastal areas are likely to be less warmer (by a very slight amount) than the rest of the study areas, because of the homogenizing effect of the sea. Thus, change is likely to be slightly lower along the Croatian coast. It might vary from 4.7°C to 4.9°C along the Albanian coast. Projected annual changes over the study area (located within the lilac block) are about 1°C higher than for those from the above-mentioned GCMs.

Winter

A very similar pattern of increase is likely for winter air temperatures (Fig. 13.5). We can expect warmer winters in the region. An increase about 4.8°C is expected in the northern part of Croatia and about 4.0°C along the coast. An average increase of

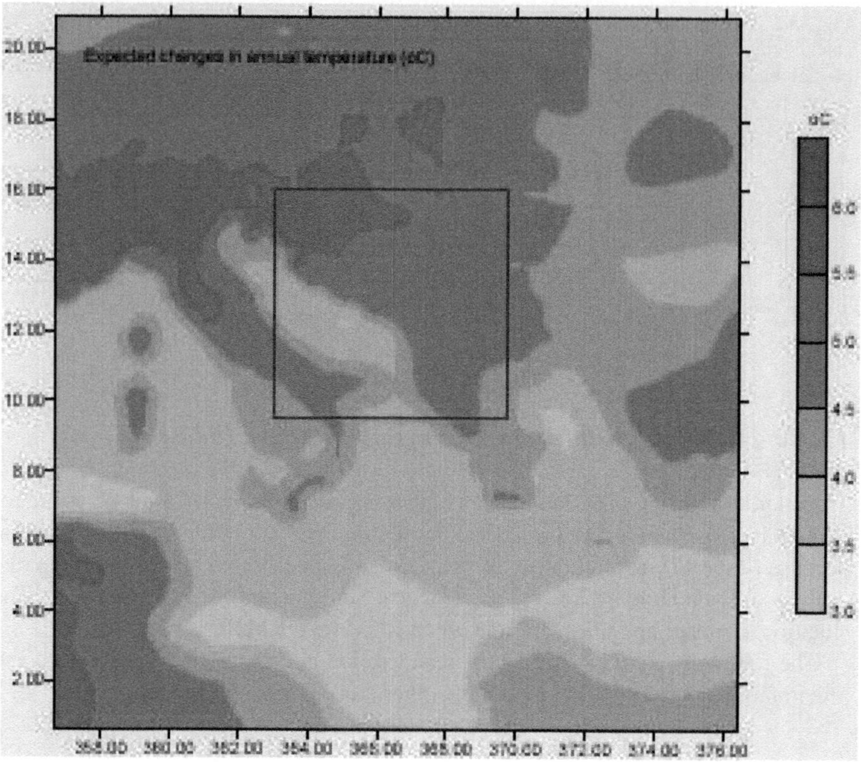

Fig. 13.4 Projected changes for the 2080s, annual temperature (°C)

about 4.5°C over the Serbian territory with slight variations moving from north (N) to south (S) (4.3–4.6 °C) or west (W) to east (E) (4.3–4.6 °C). B&H might be slightly warmer than Serbia. Almost uniform temperature changes are expected over the territories of B&H and Montenegro, 4.7 and 4.9°C respectively, with slight, non-significant changes from N to S or W to E. Winter temperatures are expected to increase about 5°C in the north and 3.8°C in the southern part of Albania. An increase about 4.5°C in winter temperatures is expected over Macedonia as well.

These changes are expected to have a strong influence on snowmelt, which will likely take place much earlier. Snow and ice are likely to decrease in mountain areas, with consequences for the timing and amount of runoff in river basins, as well as for winter tourism. As a result of increasing air temperatures in winter, the risk associated with damaging frosts will be reduced. It will allow for expansion of winter cereals and probably other winter crops.

Spring

Higher warming than in other seasonal scenarios, observed in GCMs, is expected for spring by the 2080s. Under the A2 ASF emissions scenario, HAD300 projects

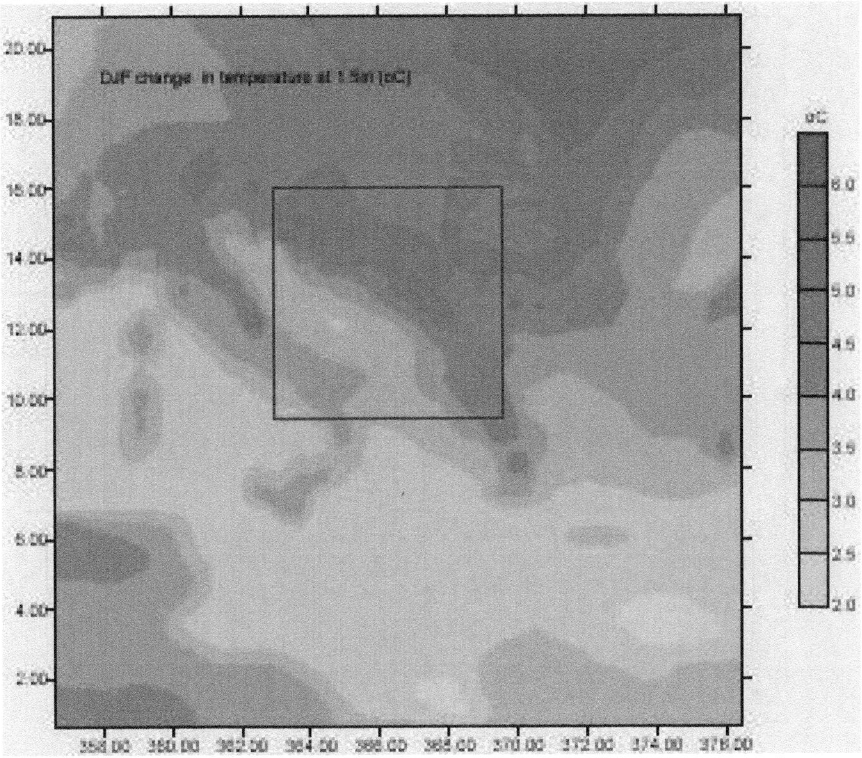

Fig. 13.5 Projected changes for the 2080s, winter temperature (°C)

increases in spring temperatures varying from 2.5 to 3.5°C. Regional model projections show – as for winter – an average increase in air temperatures of 4.3°C throughout the entire Croatian territory. This is expected to rise slightly more as one moves eastward to the northern part of Serbia (4.4°C). A warming of around 4.5°C (with very slight changes from place to place) is likely for B&H. A likely warming reaching 4.8°C is expected for southern Serbia, 5.0°C for Montenegro and up to 5.6°C for north and northeastern Albania. Inland from Albania, temperatures are projected to increase from between 3.5 and 4.6°C, with that amount increasing slightly for Macedonia, which may see changes of up to 4.8 °C.

Increasing spring temperatures will accelerate soil temperature increases and extend suitable zones for summer crops as well as lengthening their growing season.

Summer

A dramatic situation over the study area is expected for the summers of the 2080s (Fig. 13.6). Warming greater than 7.0°C is projected, varying from 7.3°C in Croatia and Montenegro to 7.5°C to east Albania on the Macedonian border. Summer values are higher than respective spring values in GSM-driven scenarios.

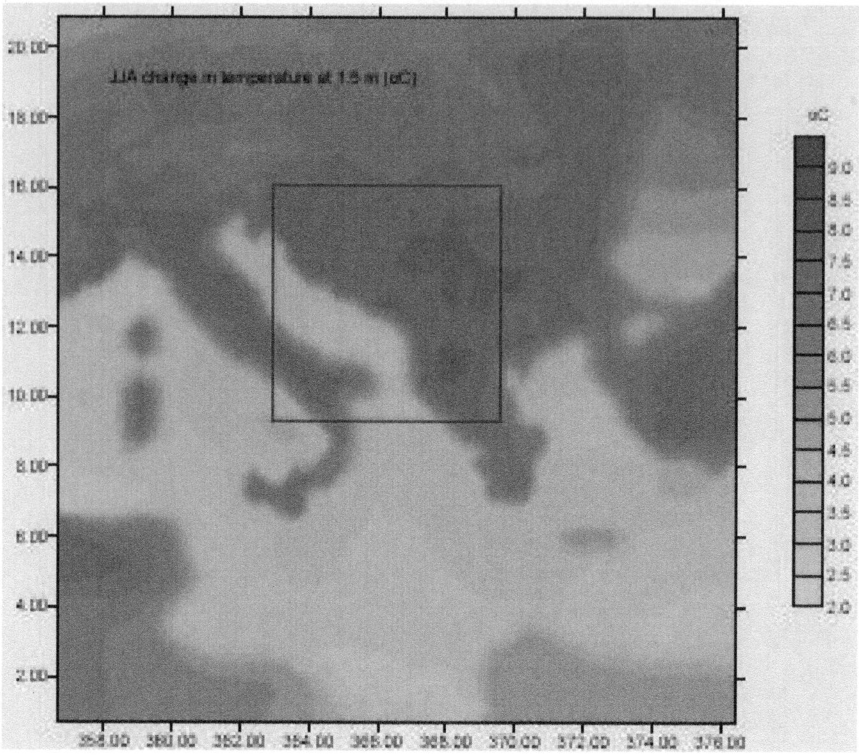

Fig. 13.6 Projected changes for the 2080s, summer temperature (°C)

Such a situation is likely to result in increases in frequency or intensity of extreme weather events (heat waves). It is known that the relationship between averages and extremes is often non-linear. For example, a shift in average temperature is likely to be associated with much more significant change on very hot days. The disproportionate increase in the frequency of extreme events is not limited to the frequency of very hot days but could occur with many other climate extremes. It will have drastic consequences on all socioeconomic systems.

Autumn

In autumn, as in other seasons, increases in temperatures are expected (Fig. 13.7), with the highest rises taking place in northern Croatia (5.2°C), northern Serbia, and B&H (around 5.0°C). In the rest of study area, with very minor differences, an increase of around 4.8°C is expected.

13.3.2.2 Total Precipitation

How will rainfall change as the climate changes? This is a key question that could have a substantial impact on society and the natural environment, as it can directly

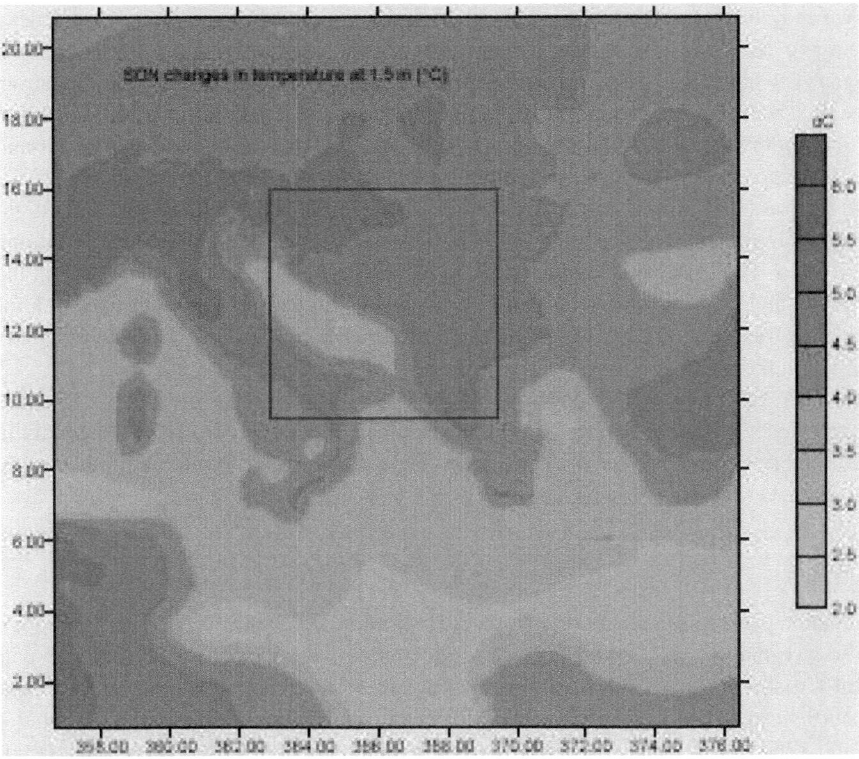

Fig. 13.7 Projected changes for the 2080s, autumn temperature (°C)

affect the availability of fresh water, the quality of potable water, and cause both drought and floods. Usually the only measure cited for rainfall is amount. Yet most of the time it does not rain. A little thought makes it clear that we also need to be concerned with how it rains: the frequency, the intensity or rainfall rate, how hard it comes down, as well as the amounts (Trenberth). Examining rainfall rates and how they change with climate change is more important and fruitful than simply examining precipitation amounts in understanding what is happening.

In fact, the general term precipitation also includes snow and other forms of precipitation in addition to rain. The patterns are very complicated. The following analysis is limited to assessment of likely changes in precipitation rates, but as they are calculated in percentages, the evaluation is valid for precipitation totals as well.

Annual Changes

It is not possible to distinguish regularity through analysis of graphs showing expected changes and gridded data of total precipitation rates.

Overall the study area is predicted to have less total precipitation for the 2080s. It must be pointed out that the pattern of changes appears to be quite irregular, thus we may expect a reduction of about 15% for western Croatia (Pazin and Rijeka areas), 16–20% for northern Croatia and 13–16% along the coast. The reduction grows to about 22% in Montenegro. Serbia is also expected to see less precipitation, with a reduction of around 7–16% in northern and 21% in central parts. The south and southwest of the country are likely to have decreasing rates of up to 18 and 21% respectively. Reduction continues through Bosnia and Herzegovina in a N–S direction (17% in the north, 20% in central regions, 23% in the south) with slight to no significant changes in the W–E direction. Expected reduction values continue to intensify for Albania (around 25% in the north, 21% in central areas, 24% in the south, and 20% on the coastal) and Macedonia (with a reduction of around 25% for the central part, varying from 20 in the north to 28% in the south and southeast, and 26% in the southwest).

As result of the reduction in annual total precipitation rates, the study area could experience a general decrease in runoff. Demands for water could rise, especially in summer. It could cause more serious problems in hydropower energy production in countries like Albania which depend heavily on hydropower.

Winter

Winter patterns of changes in total precipitation are rather different (Fig. 13.8). So, we may expect an increase in winter precipitation rates in the northern part of Croatia varying from 4 to 16% moving in a W–E direction. The northern and northwestern parts of Serbia can also expect higher rates of precipitation, about 4 and 12% respectively. It is likely to drop by about 13% in central parts of the country, 16% in the southwest and 10% in the east. A similar pattern of change will prevail in B&H. Total precipitation rates are likely to increase in the northern and northeastern parts (13%). They will start to lessen moving south, reaching negative values in the very southern part (−8%). The rest of study area is expected to see a reduction only in the total winter precipitation rate. Montenegro can expect a lower rate, around 12% over the territory, varying from 14 to 8% from N to S, and 7 to 17% from W to E. Albania will have a greater reduction in the northern mountainous area (about 20%) than in southern part (about 7%), while the Albanian coast can expect a lower rate, about 9%. Penetrating inland, the situation will deteriorate and reach values of −20% in eastern Macedonia. Moving in a N–E direction, changes in the precipitation rate will change from −6 to −26%.

It should be pointed out that with expected higher average winter temperatures, more precipitation is likely to fall in the form of rain rather than snow, which will increase both soil moisture and run off. The increase in total precipitation rates may induce soil erosion, depending on the intensity of rain episodes.

Spring

In contrast to winter, spring patterns of total precipitation rates show a decrease over the study area. The main trend is towards an intensified decrease moving in a N-S direction as well as E-W. The lowest decrease in spring rates is likely to occur

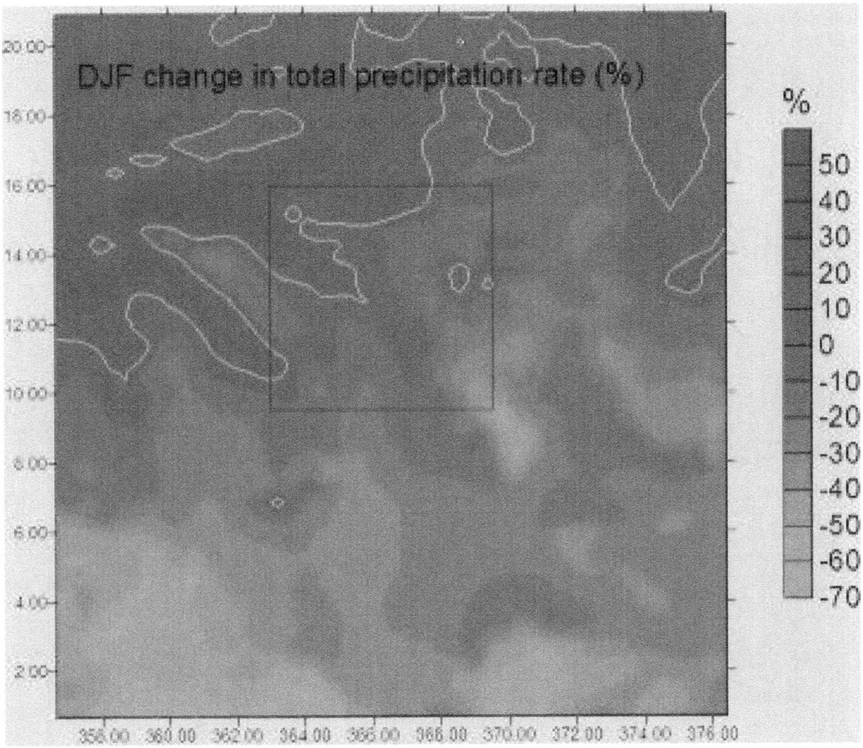

Fig. 13.8 Projected changes for the 2080s, winter total precipitation rate (%)

in northern Serbia. Moving south, such a decrease might intensify, reaching up to 16%. In the central part of the country, the reduction is expected to vary from 7% to 11 in a W-E direction. An average of about 12% less precipitation is expected in northern Croatia, and a decrease varying from 10 to 25% is likely along the Croatian coast (moving in a NW-SE direction). An average decrease of 14% over B&H (from −12% to −20% moving in a N–S direction, and −15 to −11% going from W to E) might occur in Albania, and FYROM will likely experience the highest reductions, around −27% over the territories shifting up to −30% in southern Albania.

With higher average temperatures, faster snow melt in spring is likely to aggravate springtime flooding.

Summer

Summer projections seem to be extremely problematic. Figure 13.9 shows likely changes in summer totals for the 2080s. The yellow line refers to a reduction of −50%. Thus a high decrease is expected in Croatia: around 60% in the northern part and 62–67% moving along the coast from NW to SE. A similar situation is likely to endure on Serbian territory: an average decrease of around 52%, with the highest

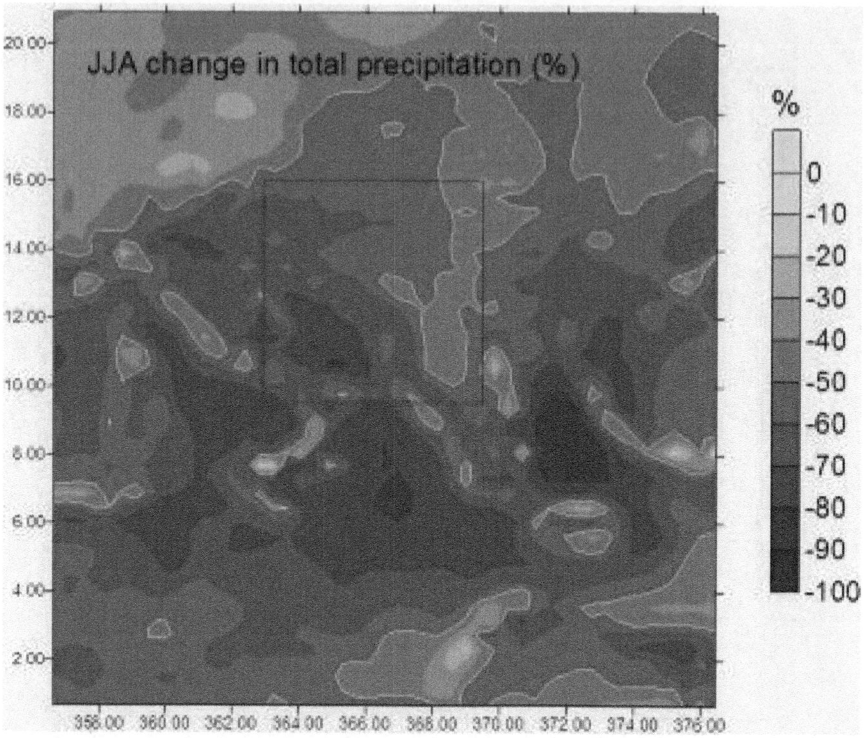

Fig. 13.9 Projected changes for the 2080s, summer total precipitation rate (%)

value W (58%) and the lowest NE (40%). Projections for B&H speak of an average decrease of around 60%, with the highest value W (64%) and the lowest S (54%). Montenegro should experience a summer with 50% less total precipitation, with slight variations over the territory. The decrease is expected in a range of 50–66% going N–S and from 64 to 46% in the W–E direction for Albania. For the FYROM, a decrease of about 50% over the territory (with slight differences) is expected.

Autumn

Expected autumn patterns of precipitation are rather different from projections for other seasons. By analyzing the data of expected changes for the 2080s, results show a very slight increase is likely for NW Croatia, but otherwise decreases varying from between 10 and 4.5% are expected N–S in the northern part and 17% along the coast. Serbia is divided into two parts, the southwest with more precipitation (an increase in precipitation rates from +1 to +7%) and the other area (N-NE-SE) with a decrease of up to 11%. Decreases (between 8 and 20%) are likely in the W-SW of B&H, followed by increases heading east (up to 2% in the NE). Montenegro can expect an average decrease of 17% over its territory, varying from 1 to 23% moving

N–S, and 23 to 10% in the W–E direction. A similar conclusion can be drawn for Albania, with an average decrease of 16% over its territory, varying in the central part from 18 to 22%. FYROM will endure an increase in its central area (from +3 to + 5%), otherwise it will see a decrease reaching up to 14% in the west and east, 3% in north and 5% in south.

It is known that increased moisture content in the atmosphere favors stronger rain and snow events, thus increasing the risk of flooding (Trenberth).

13.3.3 Impacts of Likely Changes in Climate

- There is projected to be a general drying over the study area in summer. This is ascribed to a combination of increased temperature and potential evaporation which is not balanced by increases in precipitation.
- Expected changes in surface air temperature and humidity are projected to result in increases in the heat index (which is a measure of the combined effects of temperature and moisture).
- More hot days and heat waves are very likely over nearly all the study area. These increases are projected to be greatest in areas where soil moisture decreases occur.
- Recent investigations show that increasing temperatures will be followed by an increase in the probability of extreme events and a higher intra-annual variability of minimum temperatures. More frequent and severe droughts with greater fire risk are likely.
- Increases in daily minimum more than in maximum temperatures are likely to occur over nearly all land areas.
- Frost days and cold waves are very likely to become fewer.
- Warming and population growth could increase annual heat-related deaths in those aged over 65 and contribute to the spread of vector-borne, water-borne and food-borne diseases.

Water resources

- Generally, higher temperatures lead to higher potential evaporation and decreased discharge (which also is a function of precipitation, storage, and topography). The storage of water in soil serves as a buffer; in winter and spring increasing precipitation normally generates higher discharge because the buffer is full and evaporation is low. During the summer, storage is reduced by evapotranspiration and must be refilled before discharge begins.
- As contributors to hydrological systems, snow and ice – and their potential changes in warmer global climate – will have profound impacts on streams and rivers. Higher temperatures will shift the snowline upwards and seasonal patterns of snowfall are likely to change with the snow season beginning later and ending earlier.
- Decreases in total precipitation rates combined with higher evaporative demand will probably result in less river flow (run-off).

- The hydropower industry must factor into its electricity generating considerations the probability that winter runoff is likely to increase and spring runoff probably decrease, especially in the countries dependent on hydropower (as in Albania).
- Water resources for irrigation, cities, industry and environmental flows are likely to be further stressed due to projected growth in demand and climate-driven changes affecting supply.
- Warmer average and extreme temperatures will enhance the demand for freshwater and water for irrigation purposes, especially for soils with low water-storage capacities. If precipitation declines, countries such as Albania will face a substantially increased risk of summer water shortages.

Water supply

- Numerous public supplies depend on groundwater (e.g. in Albania); any decrease in winter recharge could have serious implications. Reduced runoff would also negatively affect cooling of electric-power and industrial plants.
- Anticipated climatic changes in Albania are likely to dramatically increase the risk of summer water shortages. Significant increases in storage capacity would be needed to maintain existing water and energy supplies.
- Water quality may deteriorate because there would be less river flow to dilute contaminants (IPCC, 1997).

Agriculture

- Increasing temperatures will promote the development rate of all winter crops such as wheat, which might face extreme events and a higher intra-annual variability of minimum temperatures leading to a higher probability of crop failure from frost damage. More hot days and a decline in rainfall or irrigation could also reduce yields.
- Temperature increases in spring and summer will accelerate the course of crop development more crucially for short cycle crops sown in spring than for winter crops.
- The total growing season may be reduced for some crops. Cereal harvest dates may occur sooner. Lack of cold days could reduce vernalisation effects and consequently lengthen the first part of the growing season for winter cereals.
- Warmer winters can reduce the yield of stone fruits which require winter chilling (moderate coldness) and livestock may be adversely affected by greater heat stress.
- Summer crop yields can be affected by a shortened crop cycle and reduced time to assimilate water supplies, as well as a shorted grain-filling period. On the other hand, improvements in the rate of dry-matter production could be expected from enhanced CO_2 concentrations.
- In forestry, CO_2 benefits may be offset by decreased rainfall, increased bushfires and changes in pests.
- In cities, changes in the average climate and in sea levels will affect building design, standards and performance, energy and water demands, and coastal planning.

Energy demand

- Increases in surface air temperatures are projected to lead to increases in "cooling degree days" (a measure of the amount of cooling required on a given day once temperatures exceed a given threshold). On the other hand, warmer winters would reduce "heating degree days" and the demand for heating energy.
- Such a tendency would enhance the urban heat island effect and connected heat stress. Increased demand for irrigation water would augment the demand for energy.
- Potential direct effects of projected climate change will have further indirect effects on food security and human health.

It is imperative to integrate adaptation actions into development policy and planning at every level. This will incur incremental adaptation costs relative to plans that ignore climate change. But ignoring climate change is not a viable option – inaction will be far more costly than adaptation (Stern Review, Box 2).

Further research is needed to conduct an accurate assessment of climate impact on natural and socioeconomic fields for the SEE region. Research and development efforts over the coming decades may make current techniques more efficient or more widely applicable and may provide new techniques.

References

IPCC (2007) Climate Change 2007: The Physical Science Basis.

Trenberth KE. How should rainfall change as climate changes: Prospects for increases in extremes? National Center for Atmospheric Research. www.cgd.ucar.edu/cas/GLOB_CHANGE/extremes.html

Chapter 14
Forests in Serbia as Factor of Soil and Water Protection Against Degradation in the Conditions of Global Climate Changes

Stanimir Kostadinov

Abstract Forests in Serbia cover 27.3% of the country's total area which is considerably less than the 41.1% projected by the Spatial Plan of 1996. In addition to providing timber and other forest products, forests are increasingly important as environmental protectors and help mitigate global climate changes. Forests as well provide a powerful means to protect soil and water from all types of degradation, the primary ones being erosion and sedimentation. In Serbia, as a consequence of natural characteristics, and also due to inadequate soil and land management, soil erosion is a very significant problem, as well as the torrential floods which follow. Water erosion processes are the dominant processes in hilly and mountainous parts of the country, south of the rivers Sava and Danube, whereas in the Autonomous Province of Vojvodina in the country's north, the dominant process is wind erosion. Erosion, erosion sediment and torrential flooding cause extensive damage, increasingly negative for the environment. Problems manifest as landscape degradation, as well as mechanical and chemical pollution of water in watercourses and storage areas. Under current climate change conditions, which have an unfavourable impact on the development of vegetation, the risk of intensified erosion is increased.

This paper presents an overview of climate change in Serbia and its consequences: desertification, soil erosion, torrential flooding, as well as the significance and effect of forests in mitigating these changes.

Keywords Climate changes · Forests · Soil and water degradation · Soil erosion

14.1 Introduction

Based on current climate change discussions, including effects in various areas, most scientists worldwide have accepted the fact that human activities, primarily the burning of fossil fuels as well as systems of land use, have influenced

S. Kostadinov (✉)
Faculty of Forestry, Belgrade University, Belgrade, Serbia
e-mail: kost@eunet.rs

V. Alexandrov et al. (eds.), *Global Environmental Change: Challenges to Science and Society in Southeastern Europe*, DOI 10.1007/978-90-481-8695-2_14,

climate change. The climate is changing and will continue to change as a result of rising concentrations of greenhouse gases in the atmosphere. However, the speed and consequences of climate change in the twenty-first century are not clear, especially on a regional scale, although they will probably be serious. According to different scenarios and reports, climate change will essentially influence natural systems and key environmental resources (water resources, forests, coastal regions), and thus impact the economy. Consequently, there is great interest in effectively finding and synthesising information on the speed of future climate change and the effects of change on environmental systems. Without reliable information as a basis for further activities and decision-making, we may face considerable risks which are unfavorable for natural resources and for future generations.

Forest ecosystems belong to those natural systems likely to suffer unfavorable effects of climate changes in almost all parts of the world. During the 1970s, discussions began on the potential of forests for global climate change reduction and the importance of this potential has been accepted in numerous international discussions and analyses. Since then, there is great interest in defining and quantifying the role of forests in climate change especially when mechanisms for international cooperation have been established. Consequently, forest management will have an important role to play in climate change processes in the course of the twenty-first century. In Serbia, following the recommendations of the IPCC report, prepared according to a demand by the Science and Technology advisory organ for the UN Framework Convention on Climate Change (UNFCCC) in 2005, a 5-year programme on adapting and intensifying national activities in this field was adopted. The UNFCCC signatory countries are obliged to carry out a detailed analysis of climate changes effects on forest ecosystems and to suggest adaptation measures. The countries send their national reports to organs of the convention.

14.2 Climate Change in Serbia and the Effect of Forests

14.2.1 Temperature and Precipitation Trends in Serbia

A statistical analysis of long-term annual air temperature data series (for the period 1951–2000) shows that the territory of Serbia can be divided into two parts: in the area north of the Kragujevac–Zaječar line there is a positive temperature trend, and south of this line, there is a negative temperature trend (see Fig. 14.1). But despite global trend lines, the last two decades of the twentieth century should be taken into account because of the extremely high increase in annual air temperature registered throughout the entire territory of Serbia. The average increase in the 10-year running mean for the period 1990–2000, in relation to lowest mean values for the period 1976–1985, is about 0.6–0.8°C (or 0.4–0.5°C per decade) with a maximum increase of 1.0°C (0.6°C per decade) in eastern Serbia (Timočka Krajina) (Spasov et al., 2002).

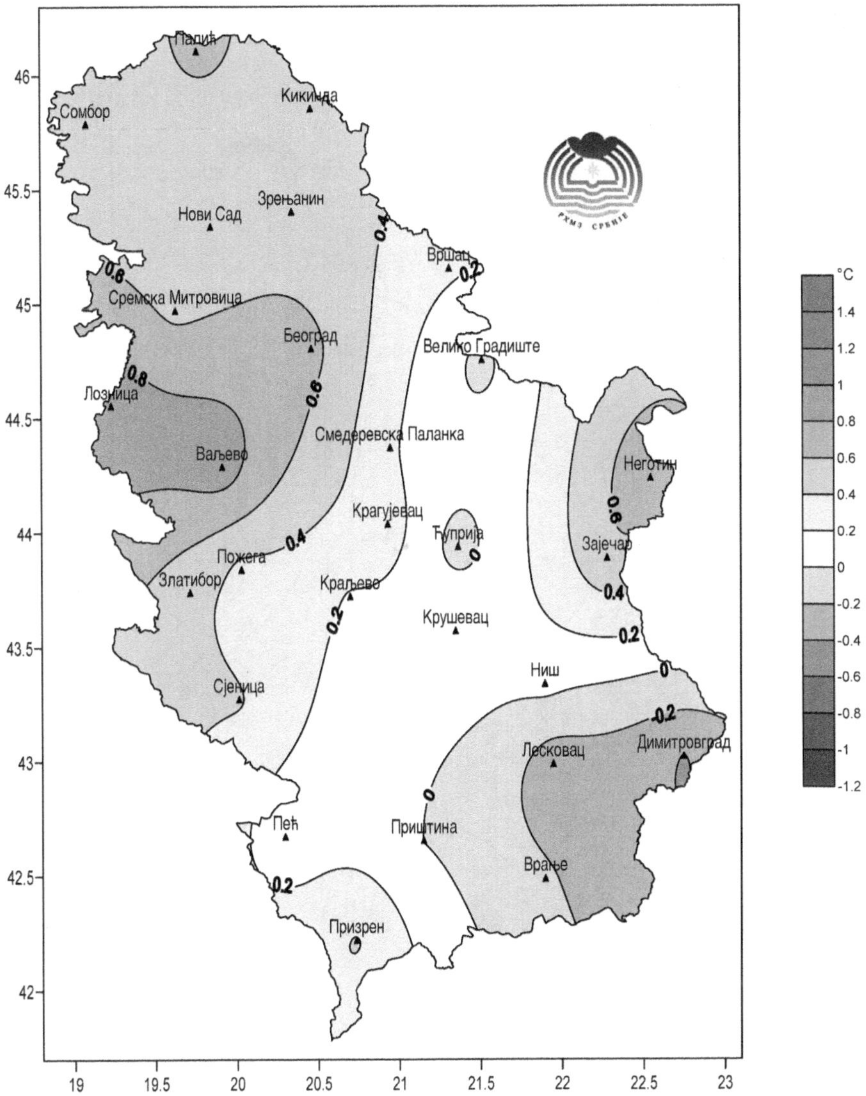

Fig. 14.1 Spatial distribution of air temperature trends (°C) in Serbia, 1950–2000

With regard to annual precipitation, results show the linear trend to be negative; this is more distinctive in the southeastern regions of Serbia, and less distinctive in the northern half of Vojvodina (see Fig. 14.2). In the utmost north (Palić), slight increases in annual precipitation totals are indicated, as well as in high altitude areas, such as Zlatibor mountain. It should be noted that at Palić and Zlatibor (two distant and different localites regarding altitude andtopographical conditions),

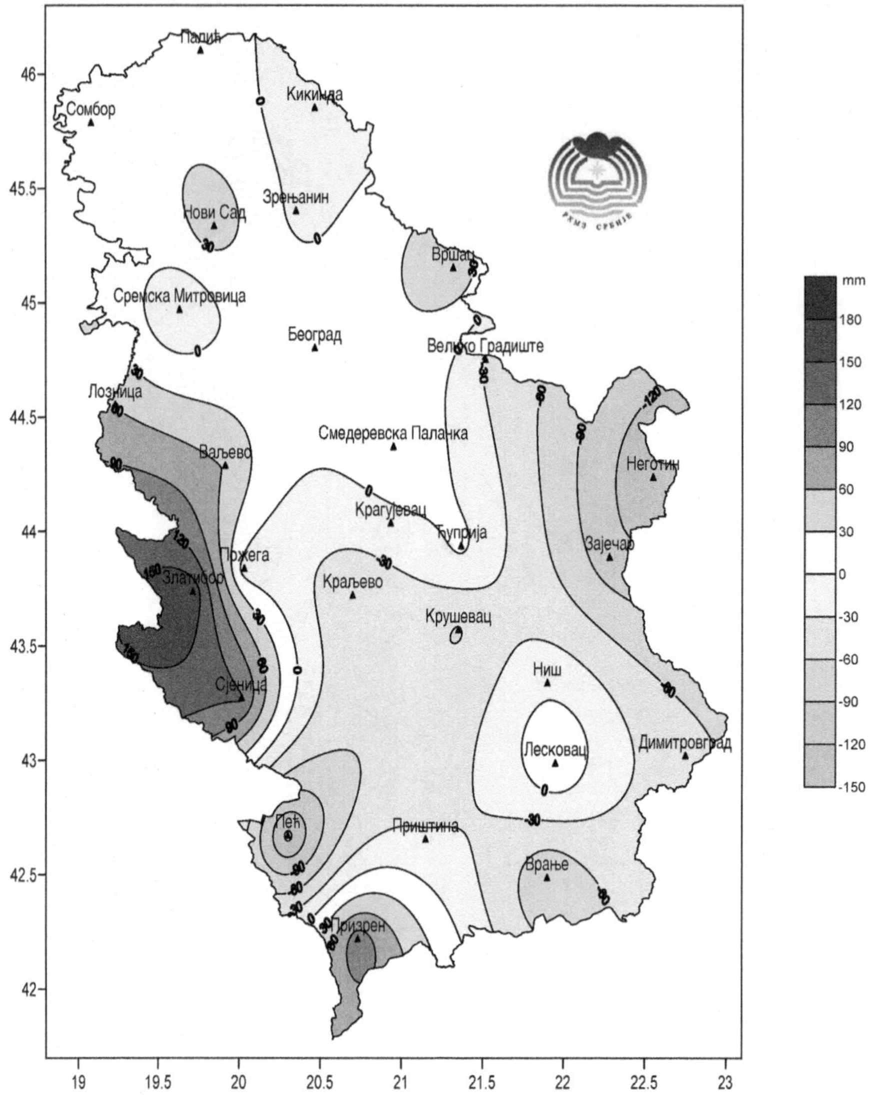

Fig. 14.2 Spatial distribution precipitation trends (mm) in Serbia, 1950–2000

minimum mean values per decade for annual precipitation are registered for the period 1956–1965, unlike for the rest of the 15 analyzed localites, where the absolute minimum is recorded for the period 1986–1995. On the other hand, during the above-mentioned 50-year period, there has been a general decrease in annual precipitation throughout the Republic (except in the higher mountainous areas of West Serbia). (Spasov et al., 2002).

Linear regression coefficients of the trend show the greatest decrease in rainfall is affecting eastern and southern Serbia, i.e. – 3.8 mm annually in Timočka Krajina and 2.2 mm annually in the south of the Republic. Rainfall has decreased over the above period by 26 and 16% respectively. The total average drop in annual precipitation for the whole territory of Serbia is about 8.6%. The greatest part of the above decrease in rainfall has taken place during the second half of the study period, especially during the last two decades of the twentieth century.

14.2.2 Forests in Serbia

The goal of afforestation in Serbia is to increase woodland areas from 27.3 to 41.4% and improve the existing situation. The basic strategy for achieving this goal is rational investment in forests which provide optimal multifunctional usage of habitation and component potentials. According to data from the National Forest Inventory of 2009, 29.6% of Serbian territory is under forest, with an increase in forested areas taking place from the period 1979 until now. Table 14.1 presents the afforested area in Serbia (Spatial Plan of Serbia 1996). Based on this data, it is clear that the percentage of forests is considerably below the optimum and the percentage of cropland – larger than what is optimal for relief conditions in Serbia. This, combined with Serbian geomorphological features, has contributed to the development of intensive water erosion in central parts of the country. The percentage of forest cover in Vojvodina is only 6%, which increases the risk of severe wind erosion. These conditions dictate that practically all of the territory of Serbia is under erosion processes of different intensities (from low to excessive erosion).

Table 14.1 Afforested area in Serbia

Area	Total area (km^2)	Area under forests 1993		Optimum afforested area (%)
		ha	%	
Republic	88.361	2,412.940	27.3	41.4
Central Serbia	55.968	1,837.417	32.8	49.8
Kosovo	10.887	429.121	39.4	52.7
Vojvodina	21.506	146.402	6.8	14.3

Source: Spatial Plan of Serbia (1996)

14.3 Soil Erosion and Sediment Transport in Serbia

Serbia is located in the central part of the Balkan Peninsula and covers 88,361 km^2. The territory consists of two different regions: the large Vojvodina plain to the north and the hilly and mountainous area to the south, with the Danube and Sava rivers representing the border between them. The geomorphological features of the Serbian territory are strongly related to soil erosion processes: wind erosion predominates in the northern plain province of Vojvodina (see Table 14.2), and water erosion in the southern region. As already mentioned, practically all of the Serbian territory is under erosion processes of different intensities. The share of certain categories of erosion intensity, i.e. erosion intensity in stream channels and in watersheds, according to S. Gavrilović's classification, is analysed below based on the erosion map prepared using S. Gavrilović's method (Gavrilović 1972) (see Table 14.3).

Sediment transport related to gross erosion (total erosion) is also considerable. The total average annual gross erosion in Serbia amounts to 37,249,975.0 m^3 and the specific annual gross erosion amounts to 421.57 $m^3 \times km^{-2}$, whereas annual sediment transport is 9,350,765.0 m^3 and specific annual sediment transport is

Table 14.2 Distribution of wind erosion in Serbia

Category	Erosion processes intensity	*Area* km²	*Area* %
I	Excessive erosion	2,888.0	3.27
II	Intensive erosion	9,138.0	10.34
III	Medium erosion	19,386.0	21.94
IV	Weak erosion	43,914.0	49.78
V	Very weak erosion	13,035.0	14.75
Total		88,361.0	100

Source: Kostadinov et al. (2006)

Table 14.3 Distribution of water erosion processes in Serbia

		Erosion categories		
	Area	Severe and excessive	Medium	Weak and very weak
Region	km²			
Vojvodina	21.506	588	3.750	10.242
Central Serbia	55.968	320	420	4.010
Central Serbia + Vojvodina	77.474	908	4.170	14.252
Kosovo and Metohija	10.887	–	285	582
Total Serbia	88.361	908	4.455	14.834

Source: Kostadinov and Spasov (2006)

105.80 $m^3 \times km^{-2}$ (Kostadinov et al., 2006). Soil erosion provokes very serious soil and water degradation.

Due to expected climate change, an increased necessity for erosion control, torrent control and terrain melioration may appear in the near future, because erosion and torrents will increase social problems which in turn will seriously impact economy, forestry and water management. Based on research results and numerous discussions worldwide, the influence of global climate change and other dynamic changes in the environment (which could be of equal if not greater importance) are pointed out. Soil degradation is not often mentioned in the context of global climate change, but it could have much worse consequences (Fig. 14.3).

Fig. 14.3 Effect of afforestation of bare land in the Vlasina river watershed

As global climate change occurs, it is almost certain that at our geographic latitudes, the following can be expected in the near future:

- increase in air temperature;
- decrease in precipitation;
- aggravation of physical characteristics of soil;
- increased tendency of soil erosion;
- decrease in vegetation's protective role;
- worsening conditions of natural and artificial revival of vegetation (Đorović and Kadović 1997).

All these changes will have a direct and indirect influence on the intensification and spread of erosion processes of all kinds. i.e., sheet erosion will considerably spread in all forms because of the increased tendency of soil erosion. A decrease in total precipitation and higher climate contrasts will increase areas potentially at risk for deep erosion, primarily because of the aggravation of infiltration ability and the loss of general macro porosity. Alongside this, erosion control and biological protection measures as well as all forms of natural and artificial revival of vegetation will be considerably hindered and expensive.

14.4 Effect of Forest Cover on Stream Flow, Soil Erosion and Sediment Transport in Mountainous Watersheds

Studies were caried out in three small experimental watersheds in western Serbia: Dubošnički Potok, Lonjinski Potok and Đurinovac Potok, situated on the territory of the community Ljubovija. All three experimental watersheds are tributaries of the river Drina and are torrential streams. Main watershed parameters are given in Table 14.4 (Kostadinov 1996). The parent rock is the same in all the three watersheds, i.e. a sandy-schistose series consisting of metamorphosed sandstones, phylites, agrilloschists and more rarely sericites, green schists, quartz breccias, quartzites and marbles. The soil in the three watersheds is acid brown soil on schists (skeletoid).

Land use is different in the three experimental watersheds. As can be seen, 70.35% of the area in the Lonjinski Potok watershed is under well-stocked forest, the number drops to 48.52% for the Dubošnički Potok watershed and to 39.5% for the Đurinovac Potok watershed. The type of cover in all the three watersheds has the

Table 14.4 Parameters of watershed areas

Parameters	Symbol	Dubošnički Potok	Lonjinski Potok	Đurinovac Potok
Watershed area	F (km^2)	1.2464	0.7656	0.544
Watershed perimetre	O (km)	5.25	3.60	3.55
Watershed length	L_{gl} (km)	2.48	1.40	1.40
Drainage density	G (km·km^{-2})	3.26	2.38	4.04
Mean elevation of the watershed area	N_{sr} (m)	487.90	363.90	299.70
Mean slope of the watershed area	I_{sr} (%)	47.24	38.87	43.59
Stream-bed slope	I_t (%)	18.37	18.94	12.63
Erosion coefficient	Z	0.56	0.34	0.49

same characteristics, namely significant areas are covered by very degraded forests transformed into thin brushwood of eastern hornbeam, ash and oak within which very strong rill erosion occurs.

As a result of different degrees of forest cover, erosion processes of different intensities develop in the watersheds. The most intensive processes of erosion occur in Dubošnički Potok (Z = 0.56, medium erosion), then in Đurinovac Potok (Z = 0.49, medium erosion) while the weakest processes of erosion occurs in Lonjinski Potok (Z = 0.34, weak erosion). *Z* denotes the coeficient of erosion in a watershed or in an erosive area, according to Gavrilović (1972).

14.4.1 Precipitation and Runoff

Table 14.5 presents annual values of rainfall and runoff (Kostadinov et al., 1998). Symbols in the table denote: P – annual precipitation depth in mm, M_Q – annual specific discharge in $l{\cdot}s^{-1}km^{-2}$, Q_{max} – maximum discharge observed in a year in m^3s^{-1}, Q_{sp} – maximum specific discharge observed in a year in $m^3s^{-1}km^{-2}$, n – number of days in a year when a stream went dry.

14.4.2 Rainfall, Discharge and Sediment Transport

Table 14.6 presents the annual values of rainfall, specific discharge and sediment transport for the period 1980–1991 (Kostadinov et al., 1997). Symbols in Table 14.6 denote: P – annual rainfall in mm, M_Q – annual specific discharge in $l{\cdot}s^{-1}km^{-2}$, M_R – annual specific transport of suspended sediment in $m^3km^{-2}yr^{-1}$, MV – annual specific transport of bedload in $m^3km^{-2}yr^{-1}$, M_G – annual specific transport of total sediment in $m^3km^{-2}{\cdot}yr$.

14.4.3 Discussion of Results

Study results show that forest cover has a considerable effect on formation of the runoff regime of a watershed. With all the other conditions being equal or similar (rainfall, relief, parent rock, soil), there is a balanced regime of runoff in Lonjinski Potok (70% of the watershed area under well-stocked forest) whereas the torrents in Đurinovac Potok (39.5% of watershed under forest) and Dubošnički Potok (48.5% under forest) have an unbalanced regime of runoff. The runoff in these two watersheds is discontinuous with large intervals of drought and discharge occurring mainly in the form of flood waves. The unbalanced regime is also confirmed by data on maximum discharge (Q_{max} and Q_{sp}). The average annual height of maximum specific discharge in Lonjinski Potok is 5.1 times lower than that in the Đurinovac Potok and 2.30 times lower than that of the Dubošnički Potok watershed.

Table 14.5 Annual characteristics of runoff

Name of watershed	Year	P(mm)	M_Q ($l \cdot s^{-1} km^{-2}$)	Q_{max} ($m^3 \cdot s^{-1}$)	Q_{sp} ($(m^3 \cdot s^{-1}) \cdot km^{-2}$)	n
Dubošnički Potok	1980	1,020.3	13.74	0.456	0.366	24
	1981	984.5	10.45	0.376	0.307	98
	1982	794.8	5.90	0.174	0.140	7
	1983	687.3	1.75	0.174	0.140	137
	1984	705.5	12.80	0.456	0.366	49
	1985	509.5	4.14	0.174	0.140	104
	1986	722.6	4.47	0.965	0.774	191
	1987	873.0	4.99	1.050	0.842	205
	1988	602.1	2.15	0.028	0.022	185
	1989	747.1	2.12	0.203	0.163	214
	Average	764.7	6.25	0.406	0.326	121.4
Lonjinski Potok	1980	1,054.7	12.44	0.073	0.095	0
	1981	911.2	10.16	0.073	0.095	0
	1982	829.6	7.78	0.055	0.072	0
	1983	768.0	5.64	0.073	0.095	5
	1984	946.1	14.58	0.360	0.470	0
	1985	651.3	8.63	0.073	0.095	5
	1986	612.2	3.76	0.040	0.052	0
	1987	995.5	11.73	0.428	0.559	0
	1988	737.1	8.26	0.073	0.095	0
	1989	875.2	9.83	0.334	0.435	5
	1990	700.0	3.52	0.028	0.036	41
	1991	828.4	6.34	0.040	0.052	5
	1992	1,020.0	7.74	0.018	0.024	41
	1993	509.0	7.70	0.040	0.052	13
	1994	463.0	2.13	0.011	0.014	99
	1995	972.0	7.26	0.040	0.052	0
	Average	804.6	7.97	0.110	0.143	13.4
Đurinovac Potok	1981	1,011.2	12.19	0.124	0.228	91
	1982	779.6	9.14	0.980	1.801	185
	1983	734.1	5.84	0.153	0.281	232
	1984	906.1	10.42	0.720	1.323	206
	1985	591.3	4.51	0.074	0.136	197
	1986	703.0	5.03	0.450	0.827	228
	1987	674.6	4.83	0.900	1.654	220
	1988	889.1	14.82	0.227	0.417	208
	1989	916.6	14.04	0.300	0.551	174
	1993	570.4	2.62	0.052	0.096	230
	1994	630.6	0.80	0.124	0.228	257
	1995	929.4	8.11	0.670	1.232	175
	Average	778.0	7.70	0.398	0.731	200.2

Study results also show that the degree of forest cover in hilly-mountainous watersheds significantly affects the type and intensity of erosion processes as well as sediment transport. The specific mean annual transport of total sediment in the

Table 14.6 Annual characteristics of rainfall and sediment transport

Name of watershed	Year	P(mm)	M_R ($m^3 \cdot km^{-2}$)	M_V ($m^3 \cdot km^{-2}$)	M_G ($m^3 \cdot km^{-2}$)
Dubošnički Potok	1980	1,020.3	57.91	225.10	283.01
	1981	984.5	81.55	30.91	112.46
	1982	794.8	76.46	34.86	111.32
	1983	687.3	28.26	14.69	42.95
	1984	705.5	254.22	52.45	306.67
	1985	509.5	44.33	256.65	300.98
	1986	722.6	32.67	33.76	66.43
	1987	873.0	316.01	464.55	780.56
	1988	602.1	3.31	0.00	3.31
	1989	747.1	106.98	57.99	164.97
	Average	764.67	100.17	117.10	217.27
Lonjinski Potok	1980	1,054.7	16.03	0.00	16.03
	1981	1,011.2	38.01	0.00	38.01
	1982	779.6	48.16	0.00	48.16
	1983	768.0	40.26	0.00	40.26
	1984	906.1	64.31	0.00	64.31
	1985	591.3	13.46	0.00	13.46
	1986	612.2	3.01	0.00	3.01
	1987	995.5	119.84	0.00	119.84
	1988	737.1	90.98	0.00	90.98
	1989	875.2	32.63	0.00	32.63
	1990	700.0	12.97	0.00	12.974
	1991	1,159.0	101.49	0.00	101.49
	Average	849.16	48.43	0.00	48.43
Đurinovac Potok	1981	1,011.2	44.27	110.98	155.25
	1982	779.6	101.47	76.82	178.29
	1983	734.1	10.78	0.00	10.78
	1984	906.1	24.64	18.83	43.47
	1985	591.3	11.87	22.31	34.18
	1986	703.0	30.85	63.53	94.38
	1987	674.6	129.44	263.26	392.70
	1988	889.1	152.38	295.46	447.84
	1989	916.6	72.75	123.34	196.09
	Average	800.62	64.27	108.28	172.55

Lonjinski Potok (70% of watershed under well-stocked forest) is 4.5 times lower than in the Dubošnički Potok (48.5% under forest) and 3.5 times lower than in the Đurinovac Potok (39.5% under forest). Such a degree of forest cover in a watershed dominates the energy potential of the watershed (relief) and the erosion activity of rainfall; this is proven by the example of the Lonjinski Potok watershed. These figures confirm that forest cover is a powerful means of flood peak control (water conservation), soil erosion intensity and sediment transport in a small mountainous watershed.

14.5 National Forest Programme in Serbia

In Serbia, afforestion of 1.1 million hectares is anticipated to be realised between 2015 and 2050. Due to the climate change, plant production must be adjusted to new conditions (i.e., a decrease in precipitation and increase in air temperature). Climate change may cause considerably larger damage to woodlands also because of strong winds, drought and forest fires in the future. For example, there were five forest fires on 12 hectares in 2003, while in 2007, 482 fires were registered, damaging an immense 34,000 ha of woodland. Fire damage affects not only wooden material and ground flora but also the structure and production abilities of soil for a long period of time. Such soil is much more prone to erosion.

Afforestation in Serbia takes place at a rate of 3,000 ha/year, but plans are in place to intensify this work by an order of three times, with a set objective of 100,000 hectares of new woodland by 2015. Afforestation and tree planting, which would protect fields, is planned for an area of 40 km^2 by 2015. Due to increased levels of environmental pollution caused by industrial and infrastructural development, planting of protective (emissions reduction) forests in industrial zones and near roads is necessary.

As part of biological protection measures and other ecological improvements, including agricultural land use, the following changes are anticipated by 2015 in Serbia (Kadović and Medarević 2007):

- afforestation of shallow, erosion prone furrows VI class quality and parts of furrows under excessive erosion II – V class quality on 200 km^2;
- afforestation of low productive grass-land IV, V, VI i VII class quality on 337 km^2;
- exclusion from production of 337 km^2 of various agricultural lands in order to afforest areas around collective basins, roads of greater importance and other sources of air pollution;
- afforestation of areas around cities, which would decrease agricultural land for 50 km^2, and active area of 36 km^2 in the form of barren and ash soil.

The expected effects of afforestation would be:

- reduced greenhouse effect because 600,000 tones of carbon would be absorbed;
- 400,000 m^3 of annual wood production or 10,000,000 euro per year;
- reduced negative effects of water and wind erosion and slides;
- reduced negative effects of polluting emmissions;
- increased agricultural production;
- better living conditions in urban areas;
- increased employment;
- significant advancement of environmental quality as a whole.

The following is part of the main goals of the Spatial Plan in Serbia and refers to water usage and protection:

- improvement of water ecosystems by protecting swamps and other water areas (natural fish spawning in riversides) by revitalizing damaged natural vegetation in these zones;
- protection from water and wind erosion in order to prevent mechanical silting of water reservoirs, mechanical and chemical pollution of water streams using technical and biological measures, in Vojvodina (from wind erosion) – by afforesting for protection purposes.

As part of an environmental protection plan, land protection is suggested, including determining agricultural land requiring re-cultivation, afforestation, erosion control as well as hydro-technical and agrotechnical melioration. Afforestation for the purposes of protection around roads, water reservoirs, settlements, industrial zones and agricultural land is anticipated as well. In order to protect flora and fauna (especially species that migrate), it is necessary to provide connections between vegetation corridors and nearby forest and swamp areas to enable migration (traffic corridors and such).

14.6 Conclusion

Woodland areas in Serbia equal 27.3% of the total territory, which is considerably less than the optimum woodland size of 41.4%. Insufficient afforestation is one of the reasons for intensive degradation of soil and water (water and wind erosion) in Serbia. Studies conducted in experimental watersheds in Serbia confirm that forests are powerful when it comes to obtaining a balanced runoff regime and successful erosion control.

Forests in Serbia, as well as in other ecosystems, are under the influence of global climate change over the last decades, and could face even greater danger according to some climate change scenarios. In order to mitigate the influence of climate change in Serbia, it is necessary to improve existing woodland, to afforest greater areas to reach optimum woodiness (41.4%) and to afforest areas around cities and industrial zones. An increase in woodland areas would reduce soil erosion, sediment transport and all damages caused by these processes to agriculture, water management and other branches of the economy.

References

Đorović M, Kadović R (1997) Perspektive i razvoj konzervacije zemljišta i voda. Uređenje, korišćenje i očuvanje zemljišta, JDPZ, Novi Sad, Srbija.

Gavrilović S (1972) Inženjering o bujičnim tokovima i eroziji (Engineering of Torrent and Erosion). Izgradnja, Special edition (in Serbian), Belgrade, 1–292.

Kadović R, Medarević M (2007) Šume i promene klime (Forests and climate changes). Proceedings of papers, Ministry of agriculture, forestry and water management of Republic of Serbia, Faculty of Forestry, Belgrade, 1–206.

Kostadinov S (1996) Soil erosion and sediment transport depending on land use in the watershed. In: Josef Krečck, K, Rajwar GS and Haigh MJ (eds) Hidrological Problems and Environmental Management in Highlands and Headwaters, Oxford and IBH Publishing CO.PVT.LTD., New Delhi, Calcuta, pp. 43–51.
Kostadinov S, Dragović N, Popović M, Nikolić M (1997) The sediment transport in the torreential watersheds as a function of the natural conditions and the erosion control works. In: Proceedings of the 3rd International Conference on the Development of Forestry and Wood Science Technology, Vol. 1, Belgrade: 168–175.
Kostadinov S, Spasov P (2006) Indicators of desertification in Serbia. In: Abstracts of BALWOIS, Conference of Water Observation and Information System for Decision Support, Ohrid, Republic of Macedonia: 10, Full paper on CD.
Kostadinov S, Vučićević S, Marković S, Nikolić M (1998) Runoff regime in small watersheds with different degree of forest cover. In: Haigh MJ, Josef Krecek K, Rajwar GS, Kilmartin MP (eds) Headwaters: Water Resources and Soil Conservation, Proceedings of Headwater 98 The Fourth International Conference on Headwater Control, A.A. Balkema, Roterdam/Brookfield, pp. 263–271.
Kostadinov S, Zlatić M, Dragović N, Gavrilović Z (2006) Soil erosion in Serbia and Montenegro. In: Boardman J, Poesen J (eds) Soil Erosion in Europe, John Wiley & Sons, Ltd., Chichester, pp. 271–277.
Spasov P, Spasov D, Petrović P (2002) Changes in drought occurrences in Serbia. In: Book of abstracts of International Conference on Drought Mitigation and prevention of Land Desertification, ICID, EWG of ICID, Bled, Slovenia: 47, Full paper on CD.

Part IV
Impacts on Society and Economy

Chapter 15
Impacts of Climate Change on Tourism

Plamen Mishev and Milkana Mochurova

Abstract This paper outlines the conceptual framework of differing climate change impacts and justifies the necessity of economic impact studies, especially at a local level. Results of a number of global impact studies (regarding international tourist flows and expenditures), as well as those conducted at both European and country levels (about both summer and winter tourism) have been examined. The purpose is not to give single values to damage or the impact of climate change, but to explore the plausible range of impacts. This paper also presents a new research project – CLAVIER (Climate Change and Viability: Impacts on Central and Eastern Europe) funded by the 6th Framework Programme and aimed at filling in the research gap concerning local level studies, especially those related to economic impact and vulnerability issues.

Keywords CLAVIER project · Climate change · Economic impact · Europe · Tourism

15.1 Introduction

A warming trend throughout Europe has been well established (+0.90°C for 1901–2005), with the recent period showing a trend considerably higher than the mean trend. Precipitation trends are variable and the number of extreme weather events has been increasing. According to the Fourth Assessment Report by the Intergovernmental Panel on Climate Change (IPCC), the wide-ranging impacts of changes in the current European climate have been documented for the first time: retreating glaciers, longer growing seasons, shifting ranges for animal species, and health impacts due to a heat wave of unprecedented magnitude. Nearly all European regions are anticipated to be negatively affected by some future impact of climate

P. Mishev (✉)
University of National and World Economy, Sofia, Bulgaria
e-mail: m.mochurova@iki.bas.bg

V. Alexandrov et al. (eds.), *Global Environmental Change: Challenges to Science and Society in Southeastern Europe*, DOI 10.1007/978-90-481-8695-2_15,

change, which will pose challenges to many economic sectors (Alcamo et al., 2007). According to research carried out by UNWTO (2007), CO_2 emissions from international tourism – including all forms of transport – accounted for less than 5% of the world total, or 1,307 million tonnes in 2005.[1] Emissions from tourism are far below those of many other sectors, such as agriculture, with 15% of global emissions. Apart from transport emissions, tourism is a relatively clean activity – one that governments around the world encourage as an alternative to heavy industry. A clean environment and favourable weather conditions are crucial to visitor satisfaction and are a fundamental factor for the development of a tourism sector. Therefore, tourism is more a victim than a vector of climate change. The purpose of this paper is to outline the current sensitivities and future impacts of climate change on tourism activities in Europe based on the results of existing studies and to present a new research project – CLAVIER.

15.2 Impacts and Global Trends

Climate impact is the consequence of climate change on natural and human systems (IPCC 2001). Depending on adaptation considerations, one can distinguish between potential and residual impacts. Impacts can generally be described quantitatively by changes in biophysical indicators (e.g., the primary productivity of an ecosystem, snow cover depth) or by socioeconomic indicators (e.g., revenues from ski tourism, effects on gross domestic product) (Füssel and Klein 2006). Potential economic impacts could be defined as a function of the exposure of human (socioeconomic) systems and their sensitivity to climatic stimuli. Most of these impacts in turn have economic costs. These economic costs of climate change are often known as the "costs of inaction", and they provide very important information for policy debate on adaptation and aid in identification of key areas of concern (EEA 2007).

Most publications are dedicated to biophysical impacts only. There are few studies on economic impact, including impacts on tourism, and these have mainly been conducted on a global or regional scale. For example, a study by Bigano et al. (2006) simulates the effects of development and climate change on tourism. The model predicts shifts in international tourist flows towards higher altitudes and latitudes. Climate change could negatively affect countries and regions that depend heavily on tourism income and could benefit places that are currently not popular with tourists. The authors use a statistical model (Hamburg Tourism Model, version 1.2.), the goal of which is to describe, with a high level of geographic breakdown, tourist behaviour in reaction to climate change, both in terms of changes in (domestic and

[1] Transport accounts for 75% of all emissions by the tourism sector, with aviation making up about 40% of all tourism transport emissions, road transport 32% and other forms of transport 3%. Accommodation represents about 21% of total tourism sector emissions. Emissions from tourism are also predicted to grow rapidly, with an increase of 152% predicted between the years 2005 and 2035 without concrete action to reduce them.

international) numbers and in terms of changes in expenditure decisions.[2] First, the authors construct a matrix of tourism flows from one country to the next. Second, they perturb this matrix with scenarios of population, income, and climate change. Third, resulting changes in average length of stay and expenditures are computed.

Currently, the OECD countries and Central and Eastern European regions dominate tourism, drawing over half the world's tourists and having only a fraction of the world's population. However, their share has been declining over the last 20 years, and will continue to do so. According to the Hamburg Tourism Model, for most of the twenty-first century, tourism will be predominantly Asian. The pattern of receipts from domestic and international tourists is different. The OECD first expands its market. The model predicts that after 2030 other regions, particularly Asia, will capture a larger share of the market. While the world aggregate number of domestic tourists hardly changes due to climate change, individual countries may face dramatic impacts that rapidly grow over time. According to the model, currently colder countries will see an increase in domestic tourism; warmer countries will see a reduction. Aggregate international tourism is expected to fall because of climate change, reaching a maximum decrease of 10% below the scenario without climate change by around 2025, and edging towards zero after that. Aggregate international tourism will fall because more tourists will stay in their home countries, particularly those from Germany and the UK, who constitute a large part of international tourism. By 2100, international arrivals may fall by up to 60% of the base value or increase by up to 220% of the base value for individual countries. Climate change increases the attractiveness of cooler countries, and reduces that of warmer ones. According to the model, world aggregate expenditures hardly change, first rising slightly and then falling slightly. The relationship between the current climate and impacts of climate change, however, is much more noisier for expenditures than for international arrivals and domestic tourists.

15.3 Expected Impacts in Europe

The Hamburg Tourism Model predicts drastic differences for countries throughout the world; a closer look at the expected impact in Europe is necessary. Continental studies, as well as those covering specific regions (mainly the Alps and the Mediterranean) show effects on both summer and winter tourism. The major effects expected are summarised in Table 15.1.

There are no specific studies covering economic impacts on the tourism sector in Bulgaria, however we can draw some general conclusions from the table above. It is quite probable that traditional mountain winter resorts at low altitudes and summer destinations at the Black sea will suffer from negative tourism trends, while new

[2] See the paper on http://papers.ssrn.com/sol3/papers.cfm?abstract_id=907454 (accessed 30 March 2009)

Table 15.1 Potential effects of climate change on economic activity in the tourism sector

Geographic location	Main climatic drivers	Expected potential impacts on economic activity	Level of confidence
Nordic regions, Eastern Europe	Rising temperature, changes in precipitation	Positive impact on tourism demand	Medium
Mediterranean regions, costal resorts	Rising temperature, changes in precipitation, sea level rise	Negative impact on tourism demand during summer	Medium
	Rising temperature in summer	Negative impact on tourism demand during summer, positive impact in spring and autumn	Medium-low
Low altitude mountain resorts	Rising temperature, changes in precipitation	Negative impact on winter tourism activities	Medium-high
High altitude mountain resorts	Rising temperature, changes in precipitation	Possible positive impact on snow-related activities	Medium

Source: Adapted from ETUC (2007) p. 22

destinations in the countryside offering alternative forms of tourism and summer mountain vacations might benefit from climate change.

The expected adverse effect on beach tourism at the Black Sea is also confirmed by the PESETA project.[3] The project makes a multi-sectoral assessment of the impacts of climate change in Europe for the 2011–2040 and 2071–2100 time horizons, first by assessing physical impacts and second by giving them a monetary value. However, the purpose of the study is not to provide single values regarding damages or impacts related to climate change, but to explore the plausible range of impacts. The tourism study aims at modelling major flows of tourism in Europe, explicitly considering the influence of climate (EC 2007b).

Coastal tourism, a dominant segment of the tourism market, has a marked seasonal and spatial concentration. In the PESETA project, climatic suitability for general summer tourism purposes is expressed using an index, the Tourism Climate Index (TCI), comprised of climate features temperature, humidity, sunshine, rain and wind. According to the TCI, some regions are seeing their climatic attractiveness improve in summer, while other regions (such as the Balkans) are facing

[3] PESETA (Projection of Economic impacts of climate change in Sectors of the European Union based on boTtom-up Analysis) – see details and preliminary results on http://peseta.jrc.es/index.html (accessed 30 March 2009).

deterioration. The TCI for the Black Sea is projected to decrease by four degrees from excellent (TCI: 80–100) to acceptable (TCI: 40–60). Project results indicate significant shifts in climatic suitability for tourism, with the belt of excellent summer conditions moving from the Mediterranean towards northern Europe. In the shoulder seasons (spring and autumn), TCI scores are generally projected to increase throughout Europe. The authors point out that the results only provide projections of climatic suitability patterns, not of the economic significance of climate change for tourism. The study is based on only one emissions scenario and one climate scenario; it does not take into account extreme weather events.

Winter tourism is a special topic of research interest. Studies show that there is a statistically significant trend in snow-cover reduction on the Alps in recent years. A number of authors are studying the impact of climate change on winter tourism, especially in Austria and Switzerland.

Breiling and Charamza (1999) have studied seasonal snow cover depths related to altitude and have developed a regionalised model for Austrian snow conditions. The snow model covers three steps. In the first step, the authors modelled the relation of snow to temperature and precipitation at each snow station under consideration. In the second step they modelled the best fit of all stations to a regional dependence on snow cover and altitude. In the third step they used a scenario for temperature and precipitation change and computed a new snow-cover depth. Then they researched the altitude at which this snow-cover depth is found today. At about 2,000 m altitude a warming of 2°C does not seem problematic concerning snow quantities. The authors predict that with this degree of warming almost half of the annual values of snow will remain in the range of 1965–1995 values. Thus, Austria could occasionally experience good winter seasons, but could not count on their regular appearance as in 1965–1995. In fact, according to the authors, it is the frequency of good and bad seasons that will determine the future of resorts. Developments in Austrian winter tourism and skiing infrastructure over the last 30 years have changed in accordance to decade temperature variation. The period 1965–1985 was a relatively cold one. At that time an expansion in the number of ski lifts occurred. The period 1985–1995 was considerably warmer. Most winter resorts had snow problems during this time, many of them serious ones. Artificial snow making became popular. Breiling and Charamza point out that just 0.8°C warming necessitated strong adaptation. The impact of 2°C of warming could leave only a few locations suitable for winter tourism and skiing – those at high altitudes.

The impact of snow-deficient winters at the end of the 1980s on the winter tourism industry in Switzerland have been examined by Koenig and Abegg (1997). The study shows that ski resorts in lower areas suffered severe consequences. The authors assess the snow-reliability of ski areas assuming a 100-day rule. This rule states that to profitably operate a ski area, snow cover sufficient for skiing (i.e. 30 cm) should last at least 100 days per season (between the first of December and the end of April). Various studies have shown that most Swiss ski areas above 1,200 m have matched the 100-day rule in the past and that a minimum altitude of 1,200 m ("line of snow-reliability") is required in order to operate a financially viable ski business under current climate conditions in Switzerland. If significant atmospheric warming were to take place (+2°C) the snowline in the Central Alps would rise by 300 m. Studies show that 85% (195 of 230) of the ski areas in Switzerland are currently snow-reliable. If climate change should occur as outlined above, the number of snow-reliable ski areas would drop to 144 (= 63%). The corresponding figures for single ski lifts are 40% (currently snow-reliable) and 9% (in the face of a 2°C increase) respectively. Even harsher consequences can be expected at the regional level. This is likely to threaten the regionally balanced economic growth in Switzerland which winter tourism has provided.

A number of studies also cover other climate-related consequences that are likely to affect tourist flows, e.g. water shortages in the southeast Mediterranean, threats to cultural heritage, biodiversity (see the summary in EEA, 2007).

15.4 CLAVIER Project

Central and Eastern European (CEE) countries face triple challenges, with ongoing economic and political transition, continuing vulnerability to environmental hazards, and longer-term impacts of global climate change. Literature reviews have shown that although there are useful studies at the European level, there is a lack of studies based on detailed regionalized climate models for CEE (incl. Bulgaria), especially those considering the economic impacts of climate change. The CLAVIER project (Climate Change and Viability: Impacts on Central and Eastern Europe) was developed with an aim to fill this research gap and thus make a contribution to the development of successful coping mechanisms for these challenges.

The CLAVIER project is supported by the European Commission's 6th Framework Programme as a three-year Specific Targeted Research Project, running from 2006 to 2009 under the thematic sub-priority "Global Change and Ecosystems". Researchers from six countries and different disciplines are investigating links between climate change and its impact on Central and Eastern Europe (CEE). Three representative countries are studied in detail: Hungary, Romania und Bulgaria.

The project is coordinated by the Max Planck Institute for Meteorology in Germany. Bulgarian partners include the University of National and World Economy and the National Institute for Meteorology and Hydrology at the Bulgarian Academy of Sciences.

CLAVIER addresses the following three scientific goals:

- Investigation of ongoing and future climate change and associated uncertainties in Central and Eastern European Countries (CEEC);
- Analyses of possible impacts of climate changes in CEEC on weather patterns and extreme events, air pollution, human health, natural ecosystems, forestry, agriculture and infrastructure, as well as water resources;

- Evaluation of the economic impact of climate changes on CEEC economies, concentrating on the four economic sectors of agriculture, tourism, energy supply and the public sector.

Within the framework of CLAVIER, ongoing and future climate change is being analysed in great detail based on existing data and climate projections in an effort to fulfil the needs of local and regional impact assessments.

One of the main objectives of the CLAVIER project is to evaluate the economic impact of climate change in Hungary, Romania and Bulgaria on tourism, agriculture, energy and the public sector. The University of National and World Economy is taking part in the WP4, with the task of investigating climate change impact on winter tourism in the resort of Borovets, and on regional development in the resort area. Interrelations at a local level between climatic parameters and social and economic variables are being studied using regression analysis, and by applying input–output methodology an assessment is being made on how impacts at the municipal and regional levels are transformed within the national economy. Adaptation possibilities regarding the public sector will also be analysed.

Partners in Romania and Hungary are also engaged in tourism impact studies. They will study the impact of climate change on tourism at Lake Balaton, winter tourism in the Carpathians and summer tourism in Constanta County.

The University of National and World Economy will also study the economic impact on agriculture and energy, including the agrarian sector in northeastern Bulgaria and nuclear power production, as well as vulnerabilities at a national level.

15.5 Conclusion

While the world aggregate number of domestic tourists will hardly change due to climate change, individual countries may face dramatic impacts that grow rapidly over time. A number of studies have shown that climate change could negatively affect countries and regions that depend heavily on incomes from tourism and could bring benefits to places that are currently not popular with tourists.

The economic and societal relevance of climate change for tourism will depend crucially on adaptation strategies. This first requires more detailed studies regarding the economic cost of climate change and the cost of adaptations at the local level (regional, municipal, tourist resort), while the second calls for the development of new policy frameworks at European[4] and regional levels. Despite the scale of the challenges, sufficient information is available to begin to act immediately.

[4] See the discussions questions raised by the Green paper on adapting to climate change in Europe (EC, 2007a).

References

Alcamo J, Moreno JM, Nováky B, Bindi M, Corobov R, Devoy RJN, Giannakopoulos C, Martin E, Olesen JE, Shvidenko A (2007) Europe Climate Change 2007: Impacts, Adaptation and vulnerability. Contribution of Working Group II to the Fourth Assessment Report of the Intergovernmental Panel on Climate Change, Parry ML, Canziani OF, Palutikof JP, van der Linden PJ, Hanson CE (eds) Cambridge University Press, Cambridge, pp. 541–580.

Bigano A, Hamilton JM, Tol RSJ (2006) The impact of climate change on domestic and international tourism: A simulation study. Fondazione Eni Enrico Mattei 86, http://papers.ssrn.com/sol3/papers.cfm?abstract_id=907454 (accessed 30 March 2009)

Breiling M, Charamza P (1999) The impact of global warming on winter tourism ad skiing; a regionalized model for Austrian snow conditions. Reg Environ Change 1 (1): 4–14

CLAVIER project (Climate Change and Viability: Impacts on Central and Eastern Europe): http://clavier-eu.org/

EC (2007a), Green paper from the Commission to the Council, the European parliament, the European Economic and Social Committee and the Committee of the Regions, Adapting to climate change in Europe – options for EU action, COM (2007) 354.

EC (2007b) Institute for Prospective Technological Studies, Joint Research Center, PESETA project, http://peseta.jrc.es/index.html (accessed 30 March 2009)

EEA (2007) European Environmental Agency, Climate change: The cost of inaction and the cost of adaptation, Technical report No 13.

ETUC (2007) European Trade Union Confederation (ETUC), Instituto Sindical de Trabajo, Ambiente y Salud (ISTAS), Social Development Agency (SDA), Syndex, Wuppertal Institute, Climate change and employment Impact on employment in the European Union-25 of climate change and CO2 emission reduction measures by 2030, Brussels.

Füssel NM, Klein R (2006) Climate change vulnerability assessments: An evolution of conceptual thinking. Clim Change 75 (3): 301–329

IPCC (2001) Climate Change 2001: Impacts, Adaptation and Vulnerability. Contribution of Working Group II to the Third Assessment Report of the Intergovernmental Panel on Climate Change, IPCC McCarthy J, Canziani O, Leary N, Dokken D, White K. (eds) Cambridge University Press, Cambridge.

Koening U, Abegg B (1997) Impacts of climate change on winter tourism in the Swiss Alps. J Sustain Tourism 5 (1): 46–58

PESETA project (Projection of Economic impacts of climate change in Sectors of the European Union based on boTtom-up Analysis): http://peseta.jrc.es/index.html (accessed 30 March 2009)

UNWTO (2007) World Tourism Organization, Tourism & climate change confronting the common challenges. UNWTO Preliminary Considerations, October 2007

Chapter 16
The Social Impact of Heavy Rains in Albania

Liri Muçaj, Eglantina Bruci, and Vangjel Mustaqi

Abstract Heavy rains in Albania are among the greatest problems currently facing the country, especially in low and coastal lands. Such zones often suffer from flooding caused not only by heavy rain but also by infrastructure mismanagement of cities located in these areas. It means that human factors are also determining damage caused by frequent inundation of coastal zones.

To analyze heavy rain and its social impact in detail, long-term observations from the Lezha station, representing the coastal northern part of the country, are considered. The heavy rainfall of 22 September 2002 over the whole territory caused substantial flooding in several areas of Albania, especially in the Lezha zone. The monthly total rainfall of 368.7 mm recorded in September 2002 has remained the highest value since 1951 for all monthly records in Albania. The recorded maximum 24-h precipitation of 219 mm is also the highest value. The threshold calculation for adverse weather phenomena identification and frequency distribution of meteorological variables are used as methodological tools for the study.

Keywords Albania · Heavy rain · Social impact

16.1 Introduction

The studied zone lies on the northern coastal part of Albania along the Adriatic Sea almost 10 m above sea level (Fig. 16.1). According to the climate classification of Albania, this area belongs to the North Mediterranean field zone (Jaho et al., 1975), characterized by dry summers and mild winters with abundant rainfall. Based on average data from all stations located in and around this zone (Jaho et al., 1975), its main climatic characteristics can be summarised as following:

L.Muçaj (✉)
Institute for Energy, Water and Environment, Tirana, Albania
e-mail: liri_mucaj@yahoo.com

V. Alexandrov et al. (eds.), *Global Environmental Change: Challenges to Science and Society in Southeastern Europe*, DOI 10.1007/978-90-481-8695-2_16,

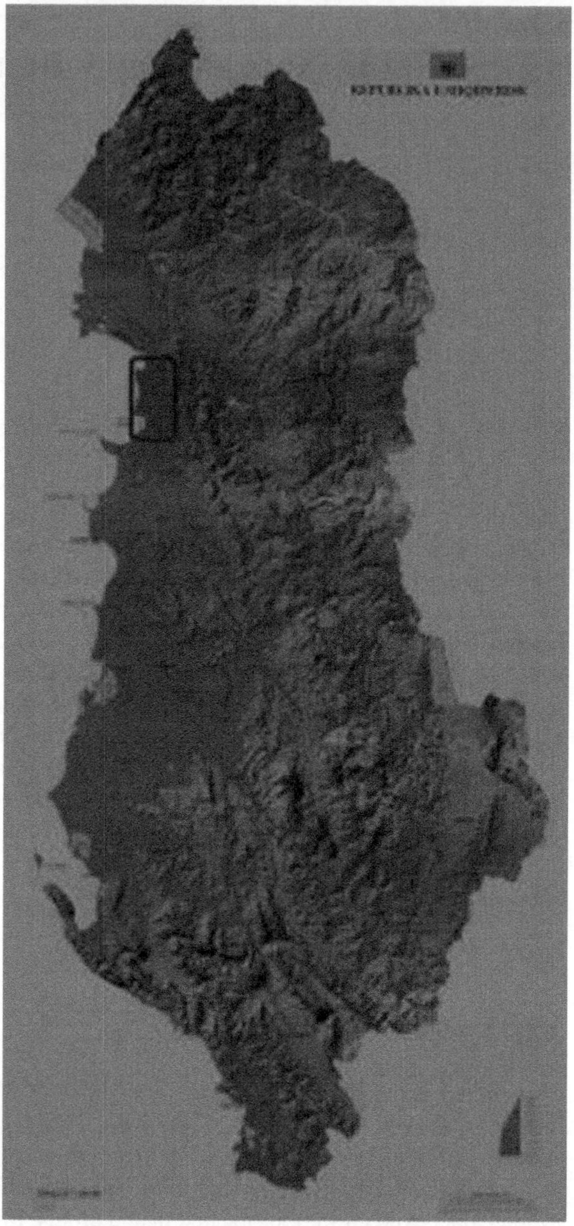

Fig. 16.1 Map of Albania

- the mean annual temperature is 15.0°C, the maximum absolute temperature – 40.4°C (4 July 1981) and the minimum absolute temperature – 10°C (24 January 1963);
- the average long-term rainfall is 1,320 mm and the recorded 24-h maximum intensity is 219 mm (21 September 2002);
- the mean wind speed is 3.7 m/s but the maximum speed reaches values of up to 40 m/s.

Frequent flooding in this zone caused by heavy rain episodes is compounded by bad management of city and area infrastructure.

16.2 Data and Methods

Heavy rain time series from 1951 to 2006 for the Lezha and surrounding stations are analysed. The data – recorded by pluviometer and pluviographs – are used to build the series showing 24 h maximum intensity. The Gumbel distribution method is applied to calculate expected values for different return periods in order to find thresholds for heavy precipitation (Radinović 1997). The function of Gumbel distribution is

$$F(x) = \exp[-\exp(-y)]$$

where $y = \alpha^* (x-u)$, x represents precipitation values, $F(x)$ the probability of overpassing the x value, and α, u are distribution parameters.

16.3 Analysis of Results

16.3.1 Annual Precipitation

The long-term average value of annual precipitation registered for the 1951–2006 period in the Lezha zone is 1,320 mm. It varies from 740 mm up to 1,900 mm (in 2003 and 1960, respectively) indicating the considerable fluctuation in rainfall totals from 1 year to the next compared to the long-term average (the normal average) (HMI 2006). In Fig. 16.2, which displays the interannual course and the long-term average of precipitation, the year 2003 may be distinguished as the driest year. The precipitation total registered for this year accounts for only 60% of the long-term average. It reconfirms the World Meteorological Organization (WMO) statement that 2003 ranks third as the driest and warmest year in Europe for the past 100 years (WMO 2003). It is also clearly evident in Fig. 16.2 that values generally remain above the mean long-term average up to 1981, after which values lower than the long-term average are registered. Annual precipitation totals show generally a trend towards decrease during the relatively short period of 1981–2006.

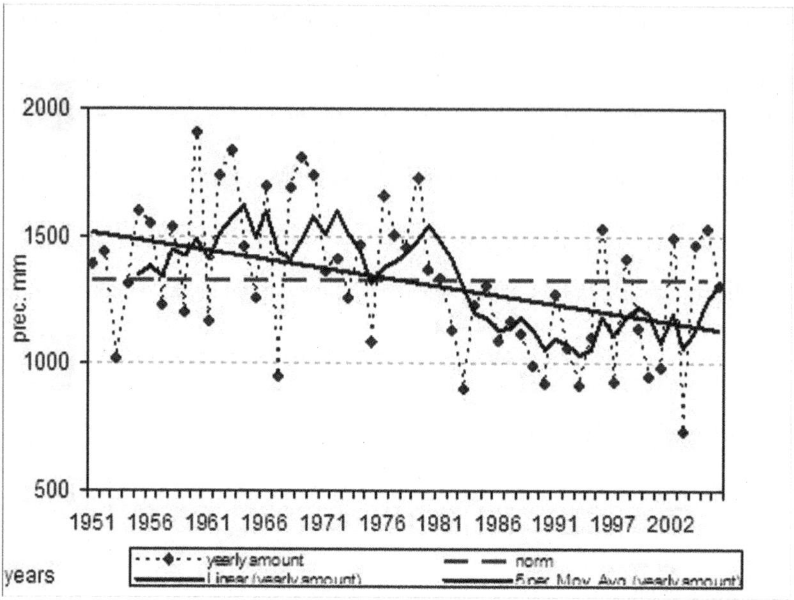

Fig. 16.2 The interannual distribution of precipitation for Lezha

16.3.2 Heavy Rain

Heavy rain can be considered a natural disaster when its intensity and frequency are far from normal, or it is widespread. This depends on the duration and level of destructiveness of the phenomenon. To calculate the threshold identifying heavy rain events, frequency distribution of the maximum value of precipitation is used. According to the level of destructiveness, adverse weather phenomena may be divided into categories as follows (Radinović 1997):

- Extraordinary (e) – adverse weather phenomena which are not destructive and only in extreme cases influence human life and human activities. The threshold above which heavy rain becomes an extraordinary event is the 24 h precipitation maximum, with a reoccurrence possibility of once in 2 years;
- Dangerous (d) – weather phenomena which directly affect human life and material goods. The threshold over which heavy rain becomes a dangerous event is the 24 h maximum of precipitation with a reoccurrence possibility of once every 5 years;
- Catastrophic (c) – weather phenomena which affect an extremely large area or reach the absolute intensity maximum at some stations. The threshold over which heavy rain becomes a catastrophic event is the 24 h maximum of precipitation combined with a reoccurrence possibility of once every 20 years.

16.3.3 September Precipitation

Before analysing the 24 h maximum precipitation for 22–23 September 2002, the distribution of monthly precipitation in September throughout the period 1951–2006 is calculated. Figure 16.3 presents the distribution of September precipitation, the long-term average precipitation (98.1 mm) and the 5-year moving average. The data series displayed in Fig. 16.3 shows that:

- there are 3 years in the September series (1972, 1998, 2002) with very high precipitation, up to 3.8 times greater than the long-term average;
- the 368.7 mm of precipitation registered in 2002 is the highest not only for this year but for the whole long-term series of September values;
- there is no significant trend, only variability, in September precipitation. The driest period seems to be 1979–1992, and the wettest ones 1971–1978 and 1993–2002;
- the lowest value of precipitation registered is 2.0 mm in 1985.

16.3.4 Rainfall on 22–23 September 2002

The highest recorded value of 368.7 mm in September 2002 was mainly due to heavy rainfall on 22 and 23 September. As a consequence of the heavy rainfall

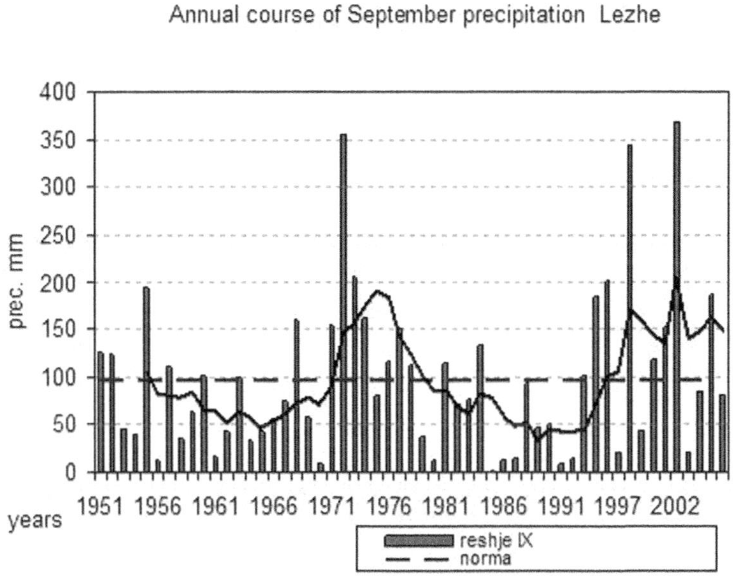

Fig. 16.3 Yearly distribution precipitation total in September for Lezha

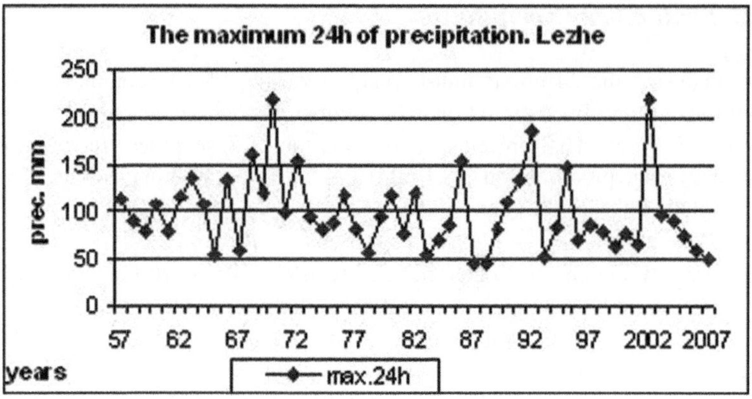

Fig. 16.4 Annual distribution of 24 h precipitation max. for Lezha

(max 24 h precipitation level up to 219 mm on 22 September), the Lezha zone was totally flooded. It should be emphasised that on this day almost the entire territory was struck by heavy rain. For example, precipitation values at various stations for that day are as follows: Koplik – 200.5 mm, Puke – 107 mm, Kukes – 92 mm, Korçe – 62 mm, Tirane – 40.5 mm, Cerrik – 52.0 mm, etc. Moreover, on this date, the 24 h absolute precipitation maximum for the territory was near to or greater than the long-term absolute maximum. The heavy rainfall caused substantial flooding not only in the Lezha zone but also in other areas of Albania. In some mountainous areas, heavy rain caused bridges to collapse, roads to be blocked, and triggered landslides, creating serious supply problems for internal and remote villages, which had trouble obtaining food and other commodities. In total, 26,000 ha of arable land were flooded and the overall loss for affected families and the country's infrastructure was estimated to be around USD 17 million, representing a very serious loss for Albania (Albanian Red Cross 2002).

Figure 16.4 shows the interannual distribution of 24 h absolute maximum precipitation for the Lezha area. It is evident that absolute maximums vary considerably year to year during the 1957–2006 period. Two years, 1970 and 2002, stand out because of very high 24 h absolute maximums of precipitation – 220 and 219 mm respectively. The lowest value of this indicator, 47 mm, is observed in 2 years – 1987 and 1988.

In order to find a rule in the annual distribution of 24-h absolute maximum precipitation in the long-term series, the respective values for 2002 are analyzed in parallel (Fig. 16.5); the monthly average of long-term precipitation is presented as well. As can be seen in this figure, there is no rule regarding monthly distribution of the 24 h absolute maximum of precipitation, which means that the 24 h absolute value could happen every month. As well, the annual course of the 24 h absolute maximum does not follow the annual course of monthly precipitation, for which the highest value of precipitation is recorded during winter and the lowest in summer.

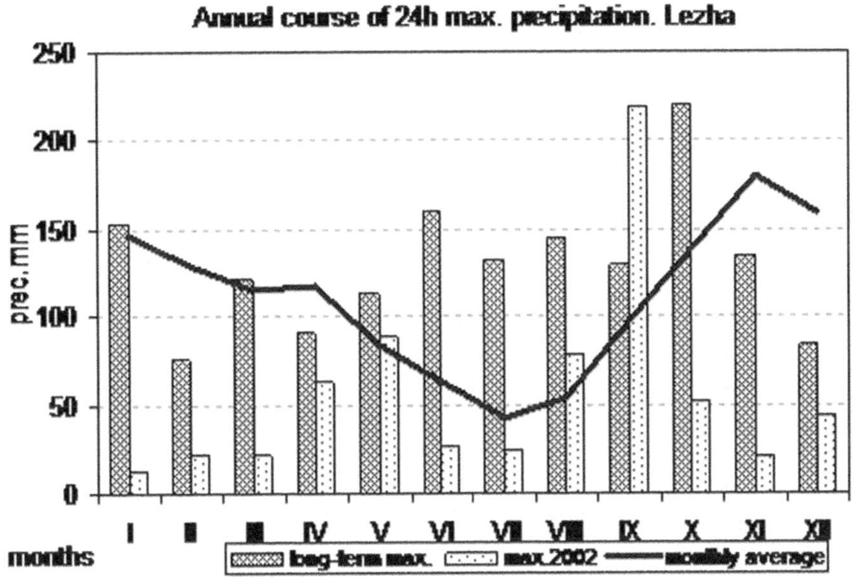

Fig. 16.5 Monthly course of 24 h absolute precipitation max. for Lezha

Moreover, it can be seen that in the long-term series, almost equal absolute maximum values of 24 h precipitation are registered for October 1970 (220.3 mm) and September 2002 (219 mm).

16.3.5 Maximum Precipitation with Different Return Periods

In the sectors of construction, hydrotechnics, agriculture, etc., the 24-h absolute maximum of precipitation, as well as maximum precipitation with different return periods are widely used parameters. The maximum intensity of precipitation within a short time interval, which is the most important characteristic of precipitation, is calculated as well (Muçaj et al. 1985). Table 16.1 presents the year-long maximum intensity and maximum intensity of precipitation on 22 September 2002. The absolute maximum intensity in the long-term series occurred on 22 October 1970. Considering the 24-h maximum intensity of precipitation for the 1957–2006 period, the greatest peaks are recorded on 22 October 1970 and 22 September 2002. Even if the difference in 24-h values between these 2 years is negligible, the value of 220.3 mm recorded on 22 October 1970 is still the highest figure. The maximum intensity of precipitation up to 12 h in duration is greater on 22 October 1970, while that for a time duration of more than 12 h is greater on 22 September 2002.

The time series of 24 h maximum precipitation for the Lezha station over the 1957–2006 period is used to find out the expected value of different return periods over a 24 h interval. Results are presented in Table 16.2. It is clear that in the study area, the expected value for the 100-year return period is 239 mm, for the 50-year

Table 16.1 Maximum intensity of precipitation for the different intervals

	Maximum intensity of precipitation									
Interval	10′	20′	30′	1 h	2 h	6 h	12 h	24 h	48 h	72 h
Total	35.2	61.6	70.4	80.3	108.5	153.0	164.1	220.3	231	242
Date	3/10/1994	3/10/1994	3/10/1994	3/10/1994	12/7/1982	22/10/1970	22/10/1970	22/10/1970	22/10/1970	22/10/1970
22/9/2002	20.2	22.5	32.5	52.0	92.5	119	144	219	258	267

Table 16.2 Expected 24-h precipitation for different periods

	Return period (year)					
T	2	5	10	20	50	100
mm	93±7	132±11	158±14	182±17	215±21	239±25

return period 215 mm, etc., which means that precipitation in the amount that fell on 22 September 2002 (219 mm) could be expected once every 54 years.

As mentioned above, the evaluation of thresholds, determining that a heavy rain is an extraordinary, dangerous or catastrophic event, is based on the 24 h maximum value of precipitation for the return period once in 2, 5 and 20 years respectively. In Table 16.2, precipitation values for categories in which heavy rain becomes a catastrophic phenomenon are presented. The thresholds are: normal situation – when the 24 h value of precipitation is below 93 mm, extraordinary – when it is more than 93 mm, dangerous – when it is more than 132 mm, and catastrophic – more than 182 mm. The heavy rains of 22 September 2002 (219 mm) thus belong to the catastrophic category. Also, based on the long-term series of 24-h precipitation, only 3 cases have been catastrophic events, 7 cases – dangerous, and 13 cases – extraordinary.

16.4 Social Impact

The Lezha area is vulnerable to flooding not only as a consequence of heavy rain but also because of bad management regarding infrastructure in the city and the surrounding area. Floods which have caused considerable economic damage occurred in January 1963, February 1966, January 1968, October 1970, January 1986, November 1992, August 1995, and September 2002 (HMI 2006). It is evident that the number of floods in the Lezha zone, especially in the last 15 years, is higher than the number of catastrophic and dangerous cases of heavy rains. As mentioned, flooding of the area is also caused by damage from hydro-technical works. For example, on 15–16 March 2000, although only 17 mm of rain fell in Lezha city and 30–40 mm over the Lezha region, the media declared that flooding was caused by heavy rain. Keeping in mind the classification of heavy rain types, it must be distinguished whether such a flood is a natural disaster or not. Human induced factors that can worsen a flooding situation include:

- reduced capacity of sucking pumps at draining stations in lowland areas;
- broken dikes in almost all affected areas;
- uncontrolled migration and settlement of people in lowland areas of the prefecture of Lezha during the past decade without proper infrastructure provided by authorities;
- electricity cuts influencing effective functioning of water pumps.

16.5 Conclusion

Results from the study presented in this chapter can be summarised as follows:

- Precipitation of 368.7 mm on 22 September 2002 in the Lezha zone is the highest not only in 2002 but over the whole long-term series for September precipitation since 1951;
- The 24 h absolute maximum intensity of precipitation – 219 mm in September 2002 – is almost equal with that of 220.3 mm in 1970;
- A heavy rainfall, such as that which occurred on 22 September 2002, has the probability to occur once every 54 years;
- The thresholds for amounts of 24 h precipitation distinguishing it as extraordinary, dangerous or catastrophic are 93, 132 and 182 mm respectively;
- Human-induced factors aggravate the flooding situation to a large degree.

Based on the above, it can be concluded that human influence is the main factor aggravating the flooding situation. Thus, local stakeholders must be responsible for proposing new development plans for the zone with a focus on city infrastructure and the energy sector, as well as for an increase in the level of public awareness regarding realization of environmental policy.

References

Albanian Red Cross (2002) Emergency appeal no. 27/02.
HMI (2006) Archives of HMI, HMI, Tirana, Albania.
Jaho S, Mukeli R, Naçi R, Mici A, Boriçi M, Bekteshi P (1975) Climate of Albania, Hidmet, Tiranë.
Mucaj L, Mustaqi V, Themelko B (1985) The Manual of Maximal Precipitation with Different Return Period, Hidmet, Tirania.
Radinović D (1997) The basic concept of the methodologies of Mediterranean cyclones and adverse weather phenomena studies. Paper presented in International symposium on cyclones and hazardous weather in Mediterranean, Palma De Mallorca (45–53).
World Meteorological Organisation (2003) WMO Bulletin 52(4).

Chapter 17
Effects of Chronic Exposure to Urban Air Pollution on Red Blood Cells in Children

Maja Nikolić and Dragana Nikić

Abstract Air quality has become a dominant factor in relation to environmental change in the last decades. The study presented in this chapter aims to indicate the significance of air quality monitoring, with special attention to children as a segment of the population at greater risk. In the study, the city of Niš is examined with regard to the effect of chronic exposure to urban air pollution on red blood cells in children. It is a pilot retrospective study including 354 schoolchildren aged 11–14 years. The group of exposed children ($n = 215$) attend school in a city area with a high level of air pollution while the children in the comparative group ($n = 139$), designated as the non-exposed group, attend school in an area with a lower level of air pollution. The diagnosis of iron deficiency anaemia is made using pre-defined criteria. The concentrations in air of black smoke, nitrogen dioxide, sulfur dioxide and lead in sediment matter are determined for the period from 1990 to 2000. The red blood cell count and average amounts of hemoglobin in blood levels of the exposed children differ significantly from those of the non-exposed children. Also, the prevalence of anaemia in children exposed to higher levels of air pollutants is significantly higher than in the non-exposed children. These findings suggest that even relatively low levels of air pollution have negative effects on red blood cells in children and increase the risk of anaemia occurrence.

Keywords Air pollution · Children · Iron deficiency anaemia · Red blood cells

17.1 Introduction

Many epidemiological studies indicate that the impact of urban air pollution on the burden of illness in the world's cities is large. In the last decade, air pollution has become a much more important topic of research in South Eastern Europe as

M. Nikolić (✉)
Faculty of Medicine, University of Niš, Niš, Serbia
e-mail: mani@junis.ni.ac.yu

V. Alexandrov et al. (eds.), *Global Environmental Change: Challenges to Science and Society in Southeastern Europe*, DOI 10.1007/978-90-481-8695-2_17,

the research into environmental changes increases. Exposure to different outdoor pollutants is associated with acute or chronic health effects, especially in sensitive members of the population (Bates 1995). Children are at high risk of developing illnesses from air pollution because of their pulmonary physiology – they have higher ventilatory rates per minute (400 mL/min/kg in newborns compared to 150 mL/min/kg in adults). Children also breathe through the mouth much more than adults and because of this the risk of pulmonary exposure to respirable airborne particles, otherwise filtered in the upper airway, increases. A higher cardiac pulse rate and the extent of tissue perfusion allow for more rapid exposure to toxins absorbed into the blood. Although air pollution has long been thought to exacerbate minor acute illnesses in children, there is increasingly strong evidence that air pollution is associated with non-trivial increases in infant mortality (Bobak and Leon 1992) and chronic disease in children (Samet and Maynard 2005).

Air pollution is a source of many substances which may enter the human bloodstream through the nose, mouth, skin and digestive tract. Most air pollutants reach the blood very quickly without previous bio-transformation, and it has been shown that they produce harmful effects on the blood, bone marrow, spleen and lymph nodes. However, investigations into the consequences of air pollution on blood have been minimal. The study presented below aims to show the importance of air quality monitoring, with special attention to children as a risk population. In the study, the effects of chronic exposure to urban air pollution on red blood cells in children from the city of Niš are examined.

17.2 Methodology

17.2.1 Study Area

The city of Niš is situated in southeastern Serbia (altitude 190 m) and covers a surface area of about 597 km^2. It is the second-largest city in the country and according to the latest census it has about 350,000 inhabitants. The whole city area is located in a valley, which is closed on three sides. The dominant wind direction is northwest and it diverges in a way which obstructs natural airing. Niš has a moderate continental climate with an average annual temperature of about 11.2°C. Temperature inversions are frequent and there are more than 100 wet days per year.

Up to the early 1990s, Niš was an industrial city. The dominant sources of air pollution have been industry and heating. During the 1990s, Niš, along with the whole country, underwent a very difficult period of economic sanction, war and poverty. The annual income per capita began to decline starting in 1989 and reached the bottom in 1993 at 1390 US$ (The World Bank Group 2008). In the same period, a great number of refugees migrated to Niš. All these factors have significantly influenced the health status of the population (UNICEF 2001).

17.2.2 Study Population

The study is comprised of children aged 11–14 years ($n = 354$), non-smokers, citizens of Niš. The group of exposed children ($n = 215$) attends school in a city area with a high level of air pollution (School 1) while the children of the comparative group, designated the non-exposed group, attend school in an area with less air pollution (School 2). School 1 is located in an urban area with major traffic roads surrounding the school building and School 2 in a residential area, far from the main street. All children live at a distance of 0.5 km from the measuring site. The exclusion criteria are any acute or chronic illnesses (to avoid bias in red blood cell results) and residence within the studied area for less than 10 years prior to the study (to guarantee significant and homogenous exposure to local air pollution).

Parents were informed about the aims, performance and expected results of the study, which took take place in the two schools. The parents agreed with involvement of their children in the study and were requested to fill in an original structured questionnaire. Data regarding demographic characteristics, parents' smoking habits, parental education levels, density of habitation (number of people living in one room) and mode of heating were collected. The procedure was approved by the Regional School Authorities of Niš (Serbian Ministry of Education).

Venous blood was analysed for hemoglobin concentrations and total red blood cell numbers at the Primary Health Care Center laboratory in Niš. The presence of anaemia is diagnosed according to the following criteria: hemoglobin of < 120 g/l and an erythrocyte count of $< 4.3 \times 10^{12}$/l.

17.2.3 Air Pollution Monitoring

The Serbian National Monitoring Network Program has conducted air quality control since 1992 in 82 sites where 20 pollutants are monitored. However, air pollution monitoring in Niš started in 1965 at two sites where daily concentrations of sulfur dioxide, black smoke and sediments were measured. During the investigated 11-year period, concentrations of black smoke, nitrogen dioxide (NO_2), sulfur dioxide (SO_2) and lead (Pb) in sediment matter were determined in 24-h samples of air collected from January 1990 to December 2000. Sampling equipment was placed 1.5 m above floor level at two sampling sites near the school buildings. Sampling sites were selected to ensure diversity regarding the outdoor environment.

The ambient level of black smoke concentrations was measured by reflection. Sampling was performed by means of a pump operating with a flow rate of 1 L/min through Whattman N°1 paper filters. At the same time, the air concentration of sulfur dioxide was measured by bubbling a certain volume of air through a solution of potassium tetrachloromercurate. Sulfur dioxide in the air stream reacts with the solution to form a stable monochlorosulfonatomercurate complex. During subsequent analysis, this complex was brought into reaction with acid-bleached pararosaniline dye and formaldehyde, yielding intensely colored pararosaniline methyl sulfonic acid. The optical density of this substance is determined spectrophotometrically at 548 nm and is directly related to the amount of collected sulfur dioxide. The total volume of air samples was determined by flow rate and sampling time. The concentration of sulfur dioxide in ambient air was calculated and expressed in $\mu g/m^3$. The lowest detectable limit is 1.7 $\mu g/m^3$.

Ambient nitrogen dioxide was collected with a pump containing triethanolamine in its tube and the exact amount of the reacted nitrogen dioxide was determined using standard spectrophotometry. The minimum detectable limit of the method is determined to be 2.0 $\mu g/m^3$. Lead in sediment matter was collected with an absorbed solution of sulfuric acid and detected by graphite furnace atomic absorption spectrometry. The lowest limit of detection is 0.5 $\mu g/m^3$.

17.2.4 Statistical Analyses

A statistical package – Statistical Package for the Social Sciences (SPSS Version 8.2, SAS Institute, Inc., Cary, North Carolina, US) was used for data analysis. Variables in hematological parameters were analysed with T-test. Air pollution data was analysed using the Mann-Whitney U-test. Statistically significant differences in anaemia incidences in children exposed to different and substantial concentrations of air pollutants were analysed using a Pearson's chi-squares test with P value of less than 0.05 required for statistical significance.

17.3 Results

The baseline characteristics of the studied population are reported in Table 17.1. There are no statistically significant differences in age between the two groups. Also, there are no statistically significant differences in parents' smoking habits, parental education levels, density of habitation and mode of heating between the two groups (Table 17.2).

Table 17.1 Distribution of children by gender and age[a]

Characteristics of children	Total ($n = 354$)	Exposed ($n = 215$)	Non-exposed ($n = 139$)
Male/female n	174/180	101/114	73/66
Age, yr (mean ±SD)	12.96±1.54	12.78±1.56	12.95±1.52
11 yr, n	109	60	49
12 yr, n	107	54	53
13 yr, n	79	55	24
14 yr, n	59	46	13

[a]No statistically significant differences between the two groups.

Table 17.2 Characteristics of examined schoolchildren

Home environment	Exposed ($n = 215$)	Non-exposed ($n = 139$)	Significance
Parental education level[b]			
Elementary	15%	31%	n.s.
Above elementary	85%	69 %	
Parental smoking habit[b]			
YES	43%	45%	n.s
Density of habitation[a] (person/room)			
Mean ± SD	0.82 ± 0.32	0.88 ± 0.29	n.s.
Wood or coal heating[b]	20.0%	15.8%	n.s.

[a]t-test.
[b]chi-square test.
n.s.– Not statistically significant.

The results of air pollution measurements are summarised in Fig. 17.1. The average concentrations of monitored outdoor air pollutants are below those in current World Health Organization guidelines. However, all air pollutant concentrations measured between 1995 and 2005 at the site near School 1 are higher compared to concentrations of the same pollutants measured at the site near School 2. This difference is statistically significant ($P < 0.05$).

The red blood cell count of the exposed children (4.16±0.32) differs significantly from that of the non-exposed group – 4.48±0.29 ($t = 9.53$; $P < 0.001$). Also, there

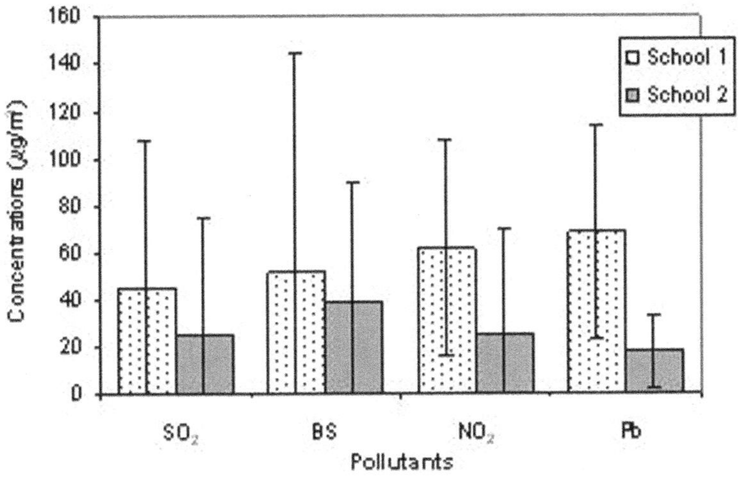

Fig. 17.1 Levels of air pollution studied between 1990 and 2000

are statistically significant differences in the red blood cell count between girls and boys of exposed and non-exposed groups (Fig. 17.2). In both sexes, the mean value of hemoglobin blood levels is significantly higher in children of the non-exposed group than those in the exposed group (Fig. 17.3).

In the group of children exposed to higher concentrations of air pollutants, 29.8% are anemic. In School 2, the percentage of childreb with anaemia is much lower (7.9 %). The chi-square test ($\chi^2 = 22.85$, $P < 0.01$) confirms that there is

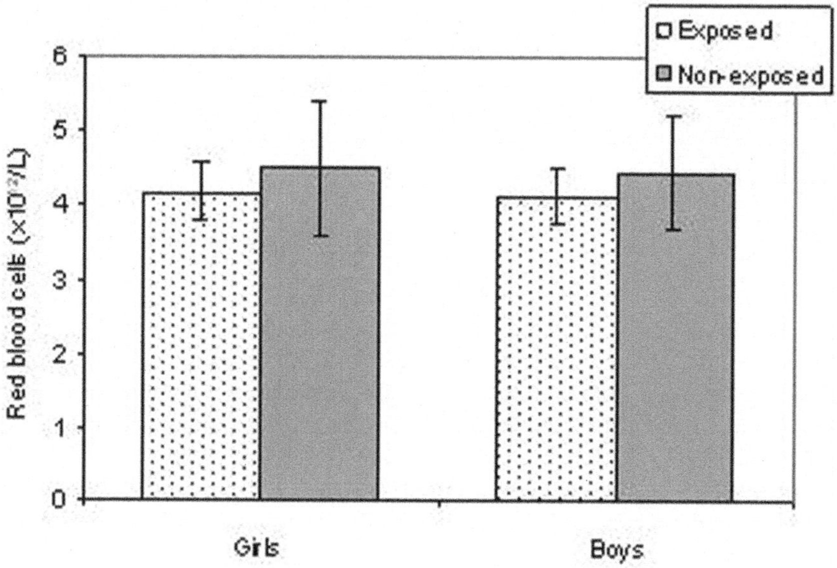

Fig. 17.2 Red blood cell count in exposed and non-exposed groups

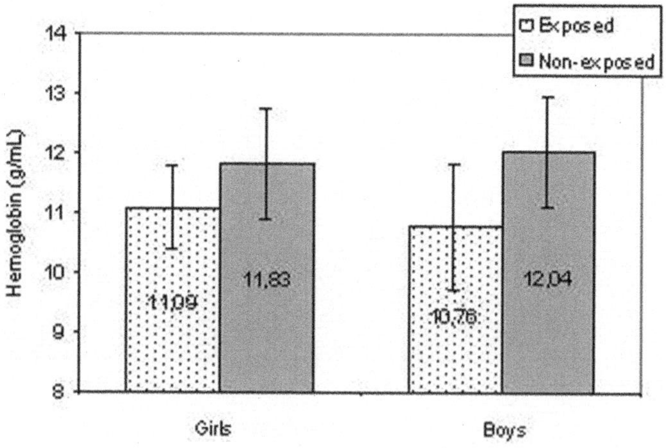

Fig. 17.3 Hemoglobin levels in exposed and non-exposed groups ($P < 0.001$)

a significant difference in the frequency of anaemia in children exposed to higher concentrations of air pollution compared to those exposed to lower concentrations. The values of relative risk are more than 1 (RR = 3.75; 95% CI: 2.06–6.88).

17.4 Discussion

Several epidemiological studies on population risk groups (pre-school children, pupils, pregnant women and elderly) were performed from 1991 to 2005 in the city of Niš, confirming that air pollution has many effects on human health (Nikić et al. 2005, 2008; Nikolić et al. 2005, 2008; Stanković et al. 2006). However, few studies investigate exposure to ambient air pollutants and its relation to anaemia. Iron deficiency anaemia, which is associated with diminished cognitive function, changes in behavior, delayed infant growth and development, decreased exercise tolerance and impaired immune function in children, is a very serious public health problem in many countries (Dallman 1982). The problem may be particularly important for poor children who are already at increased risk of delayed development (Miller 1998). Special attention should be paid to low-income and minority children in less developed countries because they face increased health risks.

The results of the investigation presented here show that even relatively low levels of air pollution, after long-term exposure, correspond positively to the occurrence of anaemia in children. Analysis of data obtained reveals a significant decrease in the number of red blood cell and hemoglobin concentrations in association with higher air pollution concentrations and long exposure.

A decrease in hematocrit and hemoglobin in association with NO_2 exposure was observed previously (Posin et al. 1978). Small but significant decreases in hemoglobin and hematocrit in adults are observed immediately after 2.5 h of exposure to 1 or 2 ppm of NO_2 in the air. Mechanisms may involve red blood cell membrane changes, methemoglobin formation, or cellular redistribution within circulation. Future studies on NO_2 exposure should consider assessment of red blood cell membranes, reticulocyte counts and methemoglobin levels. Children might be more susceptible to red blood cell effects from nitrogen dioxide exposure.

Lead poisoning is an important environmental health hazard (Bellinger et al. 1987). Highly toxic lead can also damage erythrocyte membranes, resulting in anaemia in humans. Lead interferes with many biochemical systems, particularly the heme biosynthetic pathway. Sensitivity to lead is higher in children then in adults. Lead ions inhibit enzymes that catalyze reactions for biosynthesis of hemoglobin, thus lead poisoning causes anaemia.

Although concentrations of sulfur dioxide and black smoke have been significantly decreasing over the last 10 years worldwide, the present air concentrations in the city of Niš are still an important threat to children's health. It is difficult to determine whether the measured pollutants, alone or in combinations, are solely responsible for the observed health effects in children. It is also not clear which pollutants are more responsible for anaemia in children. Little is known about possible adverse effects resulting from exposure to complex mixtures of chemicals. An analytical cross-sectional study by Ziaei et al. of 2005 shows that neonates born to mothers exposed to carbon monoxide air pollution have increased due to circulating absolute nucleated red blood cell counts compared to those of a control group. It is speculated that exposure to increased levels of ambient carbon monoxide during pregnancy may contribute to the occurrence of hypo-oxygenation. Differences in results may be due to the combined effects of air pollution and the levels of these parameters.

Further study is needed to disentangle this interaction and underlying mechanisms. Moreover, the balance of evidence remains that outdoor air pollution has a modest effect on the occurrence of anaemia in schoolchildren. Exposure to air pollution may increase the risk of anaemia in children through several mechanisms. The biological mechanisms by which air pollutants may interfere with the process of red blood cell production are reflected in the synthesis of hemoglobin, the forming of red blood cells and their life-expectancy. Toxic materials from the air can cause significant damage to red blood cells, such as reduced hemoglobin concentrations and lower numbers of erythrocytes and hematocrit, thus leading to anaemia.

Present study results are similar to previously uncovered effects of air pollution on the red blood cells of children (Nikolić et al. 2008). Naturally, other factors (genetic disposition, nutrition habits, etc.) contribute to detected low concentrations of hemoglobin and the prevalence of anaemia. The limit of this study is that only air pollution as a risk factor for anaemia has been analysed.

Air pollutants in the environment damage cell immunity and change the intensity and course of iron metabolism in the body, resulting in iron-deficient anaemia with very low hematocrit and hemoglobin values. A recent study determined the relationship between exposure to particulate matter, measured as PM10, and changes in hemoglobin concentrations, hematocrit (packed cell volume) and red blood cell count. Plasma albumin was also measured and it was found that the decrease in haemoglobin is caused by increased peripheral sequestration of red blood cells rather than generalized haemodilution. The study suggests that particulate air pollution, or a very closely associated confounding factor, has the potential to affect cardiovascular events. As mentioned earlier, children's exposure to air pollution is a special concern because their immune system and lungs are not fully developed. When exposure begins, different responses are evident in children than those seen in adults.

17.5 Conclusions

In conclusion, it can be summarized that even relatively low levels of urban air pollution have a negative impact on red blood cells in children and are hazardous to their health. Undoubtedly, a decrease in immunity and an increase in the incidence of many diseases are also connected to poverty within the children's population.

Taking into account that there are cities in Serbia with measured air pollutant levels higher than in Niš, it would be useful to conduct epidemiological studies in these cities, too. The use of unleaded gasoline has been a positive step, but alone it will not solve the problem.

This is the first study to describe the relationship between air pollution levels and anaemia associated variables among Serbian children in peer-reviewed literature. The obtained results are a baseline for further analytical epidemiological research. The correlation between red blood cell reduction and chronic high levels of air pollution is strong, but more confirmation studies are necessary. On the basis of already obtained results, it may be concluded that air quality monitoring is useful for the population groups at risk. This study may help in the design of standards for air quality in Serbia and points to the need for a reduction in limit values for pollutants considered here.

References

Bates DV (1995) The effects of air pollution on children. Environ Health Perspect 103(Suppl 6), 49–53.

Bellinger D, Leviton A, Waternaux C, Needleman H, Rabinowitz M (1987) Longitudinal analysis of prenatal and postnatal lead exposure and early cognitivedevelopment. New Engl J Med 316, 1037–1043.

Bobak M, Leon DA (1992) Air pollution and infant mortality in the Czech Republic, 1986–88. Lancet 340, 1010–1014.

Dallman PR (1982) Manifestations of iron deficiency. Semin Hematol 19, 19–30.

Miller J (1998) Developmental screening scores among preschool-aged children: the roles of poverty and child health. J Urban Health 75, 135–152.

Nikić D, Stojanović D, Nikolić M (2005) Effects of air pollution on children's health in Niš and Niška Banja. Vojnosanit Pregl 62(7–8), 537–542.

Nikić D, Bogdanović D, Nikolić M, Stanković A, Zivković N, Djordjević A (2008) Air quality monitoring in NIS (SERBIA) and health impact assessment. Environ Monit Assess, Oct 26. [Epub ahead of print] PMID: 18951135 [PubMed – as supplied by publisher].

Nikolić M, Nikić D, Stanković A (2005) Indoor air pollution in primary schools and children's health in Niš (Serbia). Cent Eur J Occup Environ Med 11(3), 189–196.

Nikolić M, Nikić D, Stanković A (2008) Effects of air pollution on red blood cells in children. *Polish J Environ Stud 17* (2), 267–271.

Posin C, Clark K, Jones MP, Patterson JV, Buckley RD, Hackney JD (1978) Nitrogen dioxide inhalation and human blood biochemistry. Arch Environ Health 33, 318–321.

Samet J, Maynard R (2005) Suspectibility of children to air pollution. In: WHO Effects of air pollution on children's health and development (A review of the evidence), European centre for environment and health, WHO, Bonn, pp. 11–13.

Stanković A, Nikić D, Nikolić M (2006) Relationship between exposure to air pollution and occurrence of anemia in pregnancy. Facta Uni Ser: Med Biol 13, 54–61.

The World Bank Group (2008) Data-country classification. (http://go.worldbank.org/K2CKM78CCO)

UNICEF (2001) Economic sanctions, health and welfare in the federal republic of Yugoslavia 1990–2000 (http://www.unicef.org/evaldatabase/index_14391.html)

Chapter 18
Moldovia's Challenge in the Face of Surface Water Resource Changes

Igor G. Sirodoev

Abstract Processes of global environmental change have had various effects upon Moldova within the last decades, both positive and negative. However, it appears that negative effects prevail, especially regarding availability of water resources during the vegetation-growing period. The impact of economic transition has had a diminishing effect on the availability of water resources. Combined with climate change trends, this double effect threatens national economy and food security, and increases the vulnerability of the population to such change. In this paper, special attention has been paid to analyzing changes in water-covered areas during the transition period in order to reveal recent tendencies and find possible adaptation solutions to the issue of climate change. "Spontaneous" concentration of reservoirs and ponds around big cities was found to be a distinct feature in recent evolutions. At the same time, in initially water scarce regions (such as in the southern part of the country) the water-covered area has diminished even more drastically (by more than 40% in extreme cases). Adaptation measures must be territorially and typologically differentiated in order to achieve maximum effect and diminish social vulnerability of the population.

Keywords Adaptation · Climate change · Republic of Moldova · Surface water resources · Water covered areas

18.1 Introduction

Assuring access to natural resources and possibilities for non-discriminatory development for future generations represents a major focus of international community activities. How these attempts are focussed depends on specific conditions in individual regions or countries. Moldova has its own particular needs in this regard.

I.G. Sirodoev (✉)
Institute of Ecology and Geography, Moldavian Academy of Sciences, Chisinau, Moldova
e-mail: ingvarri@gmail.com

V. Alexandrov et al. (eds.), *Global Environmental Change: Challenges to Science and Society in Southeastern Europe*, DOI 10.1007/978-90-481-8695-2_18,

Taking into consideration recent basic food price trends on international markets, the issue of food security should become one of the Moldova government's major concerns in the near future. Because of a lack of expensive export goods (natural resources like oil or gas, hi-tech industry products or expensive services), Moldova can only maintain its food security at an adequate level by relying on domestic food production.

Agricultural productivity in Moldova is linked to two main natural factors: soil fertility and local climate conditions. On one hand, the fertility capacity of Moldova's soils depends on many factors and is quite fragile, but it has a very strong base in the wide spreading of chernozems with a high concentration of humus.

On the other hand, Moldova's agroclimatic capacity is sensitive to variabilities in humidity. Under normal climatic conditions, non-irrigated agriculture is possible over the entire territory of the country; however, it is very vulnerable to drought. Both recent trends in weather evolution and climate projection results reveal warming and drying-up processes. This points to the possibility that droughts such as the one suffered by Moldova in 2007 – one of the most severe in the country's modern history (according to governmental assessments) – may become normal by the end of the current century.

Under these conditions, irrigation is strongly recommended to maintain agricultural productivity at a necessary level. Irrigation in Moldova was quite developed in the past, especially by the end of the Soviet period, mainly in the southeast and south. However, as a result of economic decay and transition almost all irrigation structures have been abandoned. In the mid 1990s, in only 5–6 years, irrigated areas decreased by more than 80 times, and now constitute (according to government agency "Apele Moldovei") just 1.7% of the previously irrigated area. Due to the double effect of recent and actual economic and natural processes on the availability of water for agriculture (including natural humidity and artificial watering), productivity of the country's main crops – wheat and corn – has decreased over the last years by almost 5%, At the same time, the use of water for irrigation purposes has been decreasing by 9% annually, while the volume of water used for watering every hectare of irrigated field has been decreasing by 8% per year. In addition to these dramatic data, relative water losses during irrigation have increased from 13 to 25% in the last 10 years (Sirodoev and Knight 2007b).

Territories in which irrigation structures were maintained have not suffered as much from the disastrous effects of drought. Thus extending irrigation structures must be a major consideration in order to maintain national food security at an adequate level.

Within this issue, water sources are one of the greatest concerns. The groundwater has quite a high mineralization level and requires partial demineralization before use. The country's two biggest rivers (Dniester and Prut) represent more than 90% of available water resources. However, their stream flow depends largely on climate change and it is assessed that as early as the 2020s, their water availability will diminish by 18–25% depending on greenhouse gas emission scenarios. At the same time, availability of water from local rivers will diminish even more drastically (Sirodoev and Corobov 2005; Lalikin and Sirodoev 2004).

On the other hand, climate variability and change processes in Moldova are of such a nature that using surface water in irrigation is favoured. In fact, warming and drying-up processes manifest not so much in rising average annual temperatures and diminishing annual precipitation amounts as through seasonal changes. Temperature has risen the most in winter, while other seasons have been less affected. Moreover, annual precipitation amounts remain almost the same, but their nature and seasonal distribution are changing. The warm season has become much drier, while the cold season is slightly wetter. At the same time, the proportion of extreme precipitation, such as heavy showers, is rising while the proportion of snowfall is diminishing. Future climate projections suggest these trends will be accentuated over time (Corobov and Nicolenco 2004). Under these conditions, retaining excessive runoff becomes a necessity for reasons other than just irrigation. In agriculture, capture of this runoff will be first used for redistribution of precipitation between the seasons and, in extreme cases, between wet and dry years.

Securing water resources may become essential for Moldavian agriculture, and, in a worse case scenario, lack of it could be very damaging for national economic development. Within the issue of securing water resources, maintaining and extending reservoir and pond systems will play an essential role. But how has the state of reservoirs and ponds changed?

18.2 Methods and Data

Analysis of the change in surface water resources was undertaken using change detection methodology relying on remotely sensed data. Raw data was extracted from Landsat5 imagery acquired in 1992/1993 (NASA 2003) and 2004 by pixel values. In order to avoid errors produced by shifting images of different time slices, we did not operate with absolute change values, but with "change balance", represented by the difference between an increase and a decrease in water-covered areas (lakes, reservoirs and ponds) summarized by territorial units. For the purposes of this article we used administrative-territorial units (raions).

18.3 Results and Discussions

As proven by remote-sensing analysis, territories have reacted differently to change. Several characteristics emerge from the distribution analysis of change balance (see Fig. 18.1 (left)). First of all, the presence of a certain middle axis in the distribution of growing areas is obvious. It can be explained by equidistance from the main rivers: inner regions rely primarily on local water reserves rather than on transferred supplies. That is why raions located farther from big rivers better maintain their local water storage systems, while raions neighbouring big rivers base their water supply system on in-stream withdrawal.

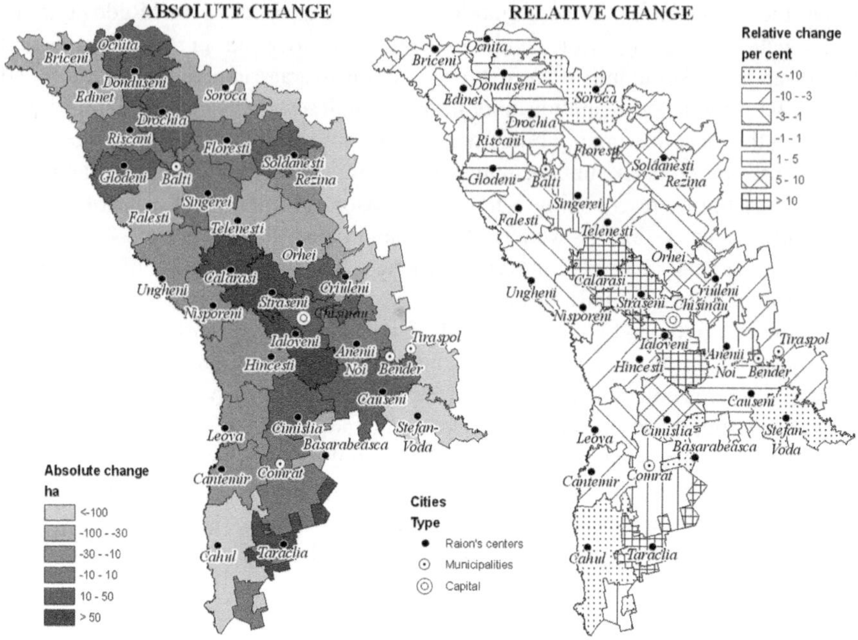

Fig. 18.1 Absolute (*left*) and relative (*right*) changes in water-covered area by raions

Three cores of positive change along the axis can be seen. Secondary cores, centred on the Taraclia raion and Balti municipality, are not very evident. High positive change in Taraclia raion is explained by an increase in the area of just one reservoir – the Taraclia reservoir. The latter is an essential part of Moldova's southern irrigation system, unfinished due to the economic crisis. The reservoir is very large and does not suggest a trend in change balance distribution.

In comparison to this, the core around Balti municipality reflects to some extent the trend of various activities concentrating near a big city. It is favoured by other factors as well (higher humidification than in other regions of the country, presence of Moldova's biggest reservoir on the Prut River etc.).

The main core of raions with increasing water-covered areas is located around Chisinau, the capital city. The area has been positively impacted by economic (man-made) factors. During the transition period, the system of reservoirs and ponds in this region was maintained and even extended due to its relationship with the capital city. Fish breeding farms, irrigated small-scale vegetable growing, tourism and recreational activities on big reservoirs, and the renting out of small reservoirs and ponds for individual use constitute the main activities which contributed to formation of this core.

Relative changes (see Fig. 18.1 (right)) in water-covered area are rooted in the same situation. Regardless of the presence of many raions with positive changes in water-covered areas, the overall balance of change is negative: since the beginning of the transition period, areas covered by natural and artificial water storage systems have decreased by 4.5% (country average). An almost total lack of maintenance work has diminished total volumes of water stored by an even greater value due to silting processes.

Positive change balance values of more than 10% have been observed in raions around Chisinau municipality. Maximum negative changes have been observed in Cahul raion due to the deliberate drying out of several big reservoirs of the Cahul fish farm complex; the Basarabeasca raion has suffered a similar fate. Water-covered areas have diminished in these raions in the transition period by about 40% each.

When comparing this situation with water scarcity vulnerability (see Fig. 18.2), the effect of changes in water-covered areas is potentially very threatening for

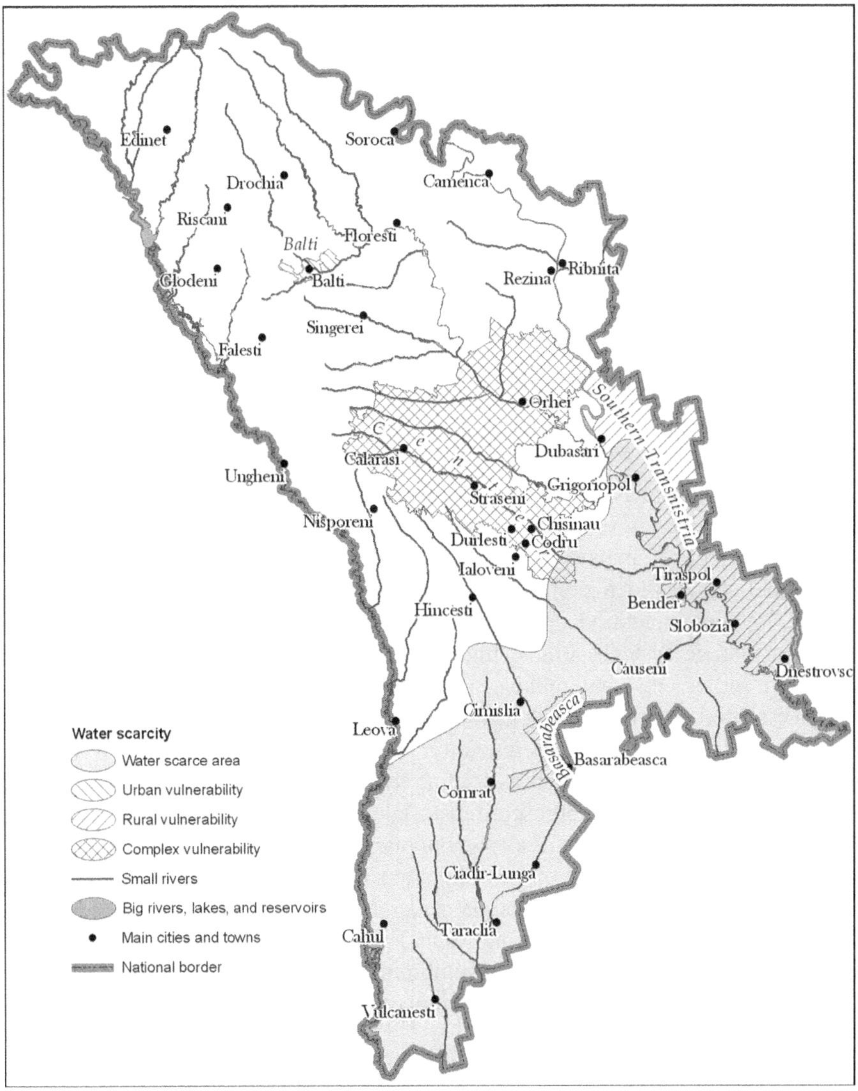

Fig. 18.2 Potential vulnerability to water scarcity (Sirodoev and Knight 2007a)

Moldova. The three raions with the biggest negative changes (Cahul, Basarabeasca, and Stefan Voda) are located in a naturally water-deficient region. Moreover, Basarabeasca region has a high rural vulnerability to water scarcity (Sirodoev and Knight 2007a). The natural deficit of water and high anthropogenic load on water resources in the southern part of the country combine with very high negative changes in stored water volume.

At the same time, territories with complex vulnerability to water scarcity are not as exposed to the negative effects of climate change as appears at first glance, analyzing the static situation of water storage and supply systems. Positive changes in water-covered areas in the raions around Chisinau municipality will help to moderate the negative effects of expected climate change. The northeastern part of the central region is more exposed because the water-covered area has diminished in the Orhei raion by 2.2%.

Regardless of some positive aspects of water storage system dynamics, the overall situation is not very promising. Disregarding maintenance work in reservoirs and ponds, the lack of a strategy for extending water storage and irrigation systems will contribute to diminishing availability of water for direct use, as well as to an overall decrease in water resources.

Shifting to an analog approach, as was done in Bulgaria (Knight et al. 2004), can lead to extremes becoming the norm in the near future. If the situation evolves according to a "business-as-usual" scenario, Moldova's water resources will deplete very quickly. In order to anticipate these effects, special attention must be paid to hydrotechnical works and adaptation measures now. Moldova has strong potential to annihilate negative effects by implementing adequate adaptation measures and through the wise and effective use of water resources and storage capacities.

Generally, several aspects could influence the success of implemented measures. Among negative ones, emphasis should be placed on natural water scarcity in the southern part of the country and the unfavourable dynamics of recent variability trends and changes in regional climate. Two positive features could help, if used wisely, in implementation of adaptation measures: "spontaneous" concentration of water storage systems around big cities (and, generally, around economically important settlements), and relative stability of annual natural precipitation values.

An essential part of the implementation strategy for adaptation measures should be dedicated to territorial differentiation of proposed measures. In our opinion, the latter should be classified according to social effect. And implementation strategies should, in fact, represent territorial and typological distribution in order to balance physiographic and socioeconomic conditions and thus achieve better results. Proposed classification of adaptation measures (passive, socially passive and socially active types) takes into consideration their social effect (Sirodoev and Knight 2007b). It is obvious that the southern part of the country should enjoy the most complex adaptation measures, combining various types with differing effects on population in order to achieve maximum effect and to minimize social instability among the affected population.

As we can deduce from analysis of the situation, the transition period created a change in the hierarchy of reasons for developing and maintaining water storage systems. In the Soviet period, irrigation of vegetables and orchards as well as flood protection of agricultural lands and settlements constituted the main reasons for extending such systems. After more than 15 years of transition, the keeping of reservoirs and ponds has become possible because of activities linked to big cities. Cities, in their relation to water storage systems, play the role of consumer of goods. Thus the situation has become one in which water storage systems are modelled on a settlement system. We can presuppose that appropriate intervention in settlement systems can accordingly bring positive effects to water storage systems.

18.4 Conclusions and Recommendations

Our studies have revealed that Moldova's water resources and food security are threatened by both "natural" changes in regional climate and the effects of the past 17 years of economic transition. It is difficult to estimate the impact of each factor at present. What is true is that both have contributed to decreasing productivity in agriculture and the growing vulnerability of national food security. The main challenge to Moldova is finding appropriate adaptation schema.

Negative climate variability and change effects reached a zenith in 2007, when one of the most severe droughts occurred. However, their influence could have been easily mitigated by appropriate management. The easiest way to adapt to expected climate change is this: hold on to long-lasting Soviet traditions regarding water use and availability of necessary resources. Regardless of the water use strategy to be implemented, appropriate management of available water resources must be of the highest priority. The change in surface water resources challenges Moldova to properly use its actual potential before seeking and implementing new, more sophisticated, and more expensive approaches to assure food security and adaptation to climate change.

Acknowledgements The research in this publication was made possible in part by Award No. MG2-2318-CH-02 of the US Civilian Research & Development Foundation for the Independent States of the Former Soviet Union (CRDF) and by Award No. MTFP-04-09 of the Moldovan Research and Development Association (MRDA) under funding from CRDF.

Any opinions, findings and conclusions or recommendations expressed in this material are those of the authors and do not necessarily reflect those of the MRDA or CRDF.

The author would like to give special thanks to Prof. C. Gregory Knight of Pennsylvania State University for the constant research help and support.

References

Corobov RM, Nicolenco AV (2004) Novye proektzii antropogennogo izmenenia klimata Moldovy v XXI stoletii (New projections of anthropogenic change of Moldova's climate in XXI century). In: Corobov RM (ed) Moldova's Climate in XXI Century: Projections of Changes, Impacts, Responses, Moldova Academy of Sciences, Chisinau, pp. 54–97.

Knight CG, Raev I, Staneva MP (eds) (2004) Drought in Bulgaria: A Contemporary Analog for Climate Change, Ashgate, Burlington.

Lalikin NV, Sirodoev IG (2004) Nekotorye podhody k otzenke vozdeystviy izmenenia i izmenchivosti klimata na vodnye resursy (Some approaches to the assessment of climate change and variability impact on water resources). In: Corobov RM (ed) Moldova's Climate in XXI Century: Projections of Changes, Impacts, Responses, Moldova Academy of Sciences, Chisinau, pp. 176–212.

NASA Landsat Program (2003) Landsat TM, USGS, Sioux Falls. Source for this data set was the Global Land Cover Facility: www.landcover.org

Sirodoev I, Corobov R (2005) An approach to the modeling of some river water quality parameters. In: The third international conference 'Ecological chemistry'. Latest advances. Book of Proceedings. Chisinau, pp. 242–248.

Sirodoev IG, Knight CG (2007a) Vulnerability to water scarcity in Moldova: Identification of the regions. Buletinul Academiei de Ştiinţe a Moldovei. Ştiinţele vieţii 3(303):159–166.

Sirodoev IG, Knight CG (2007b) Vulnerability to water scarcity in Moldova: Likely threats for future development. Present Environ Sustain Dev 2:3–15.

Chapter 19
Long-Term Trends of Lead-210 Concentrations in the Ground-Level Air in Finland and Bulgaria

Jussi Paatero, Blagorodka Veleva, and Juha Hatakka

Abstract Long-term records of airborne lead-210 should reveal periodic changes in the atmosphere, such as the North Atlantic Oscillation, as well as any systematic trends in large-scale weather phenomena transporting arctic, maritime or continental air masses which might be indications of a changing climate. The results of collaborative research on these matters show that starting from the mid-1980s, there has been a continuous decrease in average lead-210 contents in the ground-level air both in Finland and in Bulgaria. However, the decreasing trend is much stronger in Bulgaria compared to Finland. This trend may indicate that the climate is becoming more maritime in both northeastern and southeastern Europe, but that the effect is much stronger in the southeast of the continent. Because the sources of air masses are not the only factor determining lead-210 content in the air, the effect of precipitation causing wet deposition of lead-210 must be studied, too, as changes in precipitation amounts can influence the average residence time of lead-210 in the air.

Keywords Atmospheric radioactivity · Bulgaria · Finland · Lead-210 tracer

19.1 Introduction

Lead-210 is formed in the atmosphere by radioactive noble gas radon-222 emanating from the Earth's crust. Ninety-nine percent of airborne radon originates from land and only 1% from the sea (Baskaran et al. 1993). The amount of lead-210 in the air is not affected by anthropogenic activities (Hötzl and Winkler 1987). The practically exclusive formation mechanism of airborne lead-210 facilitates its use as a tracer for air masses of a continental or maritime origin. Long-term records of airborne lead-210 should reveal periodic changes in the atmosphere such as the North Atlantic Oscillation. It should also reveal any systematic trends in large-scale

J. Paatero (✉)
Finnish Meteorological Institute, Helsinki, Finland
e-mail: jussi.paatero@fmi.fi

V. Alexandrov et al. (eds.), *Global Environmental Change: Challenges to Science and Society in Southeastern Europe*, DOI 10.1007/978-90-481-8695-2_19,

weather phenomena transporting arctic, maritime or continental air masses which might be indications of a changing climate.

Both in Finland and Bulgaria, the national weather services responded in the late 1950s to the imminent need to monitor airborne radioactivity caused by intense nuclear weapons testing mainly in the United States and the Soviet Union. Soon both institutes, the Finnish Meteorological Institute (FMI) in Finland and the National Institute of Meteorology and Hydrology (NIMH) in Bulgaria, adapted monitoring methods for airborne total beta activity, because these methods are relatively simple and cost-efficient. Both institutes have been monitoring total beta activity through daily sampling. Three to five days after the end of sampling, short-lived radon-222 daughter nuclides decay into lead-210 and radon-220 progeny decay into stable lead. Thus, in the absence of a significant amount of artificial radionuclides, the measured total beta activity consists mainly of lead-210 and especially its daughter nuclide bismuth-210 (Paatero et al. 2007). In 1967, the FMI also started monitoring of daily activity concentrations of lead-210 in the air. The present analysis is based on alpha counting of in-grown daughter nuclide polonium-210 6 months after sampling.

The last atmospheric nuclear test was conducted in China in October 1980. Since 1982, the observed total airborne beta activity in Finland and Bulgaria has consisted mainly of lead-210/bismuth-210, with a notable exception being the contamination in 1986–1987 from the Chernobyl accident in Ukraine. A long-term data set of airborne lead-210 in Sofia has also been produced as part of the collaborative research, based on measured total beta activity/lead-210 ratio in a set of aerosol filters collected in NIMH.

19.2 Experimental Methods

The Finnish Meteorological Institute has collected daily aerosol samples at the Nurmijärvi geophysical observatory (60°30′N, 24°39′E, height 105 m above sea level [a.s.l.]) and at the Arctic Research Centre in Sodankylä (67°22′N, 26°39′E, height 198 m a.s.l.) since the early 1960s. The samples have been collected with a high-volume sampler and put onto glass-fibre filters with a diameter of 24 cm. The filter type was changed from Whatman GF/A to Munktell MGA in August 1995. The sampler has a capacity of 3,500 m^3 per day and collects particles with an aerodynamic diameter of less than 10–15 μm, depending on wind speed. Air flow through the filter is measured using a rotameter. The filters are changed every morning at 06 UTC (Paatero and Hatakka 2002).

The ^{210}Pb contents of samples are analyzed 6 months after they are taken by counting alpha particles from in-grown daughter nuclide ^{210}Po. Earlier, measurements were performed manually using a large-area ZnS scintillation counter. Counts from the detector were fed into an automatic logger that issued the results on punched tapes for further computer analyses. Since 1982, measurements have been carried out using two successive automatic alpha/beta analyzers, which are microcomputer-controlled and can handle 25 air filters at one loading. The detector arrangement of the analyzers consists of five flat large-area gas-flow proportional counters, alpha and beta sensitive and background shield counters above the sample, and beta sensitive and background shield counters below the sample. The flow gas is P-10, a mixture of argon (90%) and methane (10%). A constant gas flow rate is maintained with a mass-flow controller. Background samples (unexposed filters) and reference samples (^{242}Pu, ^{90}Sr and ^{55}Fe) are measured daily (Mattsson et al. 1996). Samples are archived after the measurements and are thus available for further analysis. Atmospheric pollutants, such as sulphur compounds and heavy metals, have been analyzed from these filters (Mattsson and Jaakkola 1979; Yli-Tuomi et al. 2003) (Fig. 19.1).

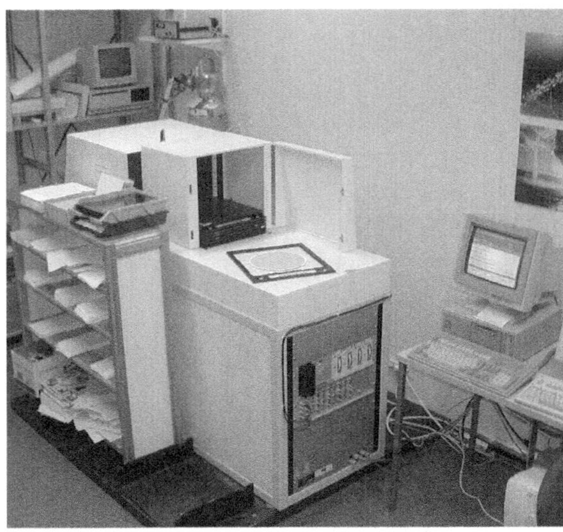

Fig. 19.1 FMI's automatic alpha/beta analyzer "Genera T84"

In Sofia (42°39′N, 23°23′E, height 589 m a.s.l.), measurements of atmospheric radioactivity started at the NIMH in 1958. Total aerosol beta activity was initially measured 3 days after the end of sampling, with that changing to 5 days since 1965. Aerosol samples are collected 2 m above a grass surface on paper filters (FPP, Synpor) with a pump. Air volume is measured using a flow meter. Samples are taken every day at 06 UTC. The air volume of samples has changed over the years from 10–15 m^3 at the beginning of measuring to the current figure of 100–120 m^3. Radiation detectors changed from GM counters to proportional counters and, at the beginning of the 1980s, to plastic scintillator detectors (Manolov and Teneva 1964).

The relationship between total beta activity concentration and ^{210}Pb activity concentration is obtained using a comparative measurement of daily filters collected in Sofia over 1 year. The filters that have been measured for total beta activity in NIMH's laboratory are assayed for ^{210}Pb in FMI's laboratory. The observed relationship is:

$$A_{\text{lead}-210}\ [\mu\text{Bq/m}^3] = (A_{\text{total\_beta}}[\text{mBq/m}^3] - 1.257)/4.476 * 1{,}000. \quad (1)$$

Yearly average ^{210}Pb activity concentration values in Sofia are calculated from yearly average total beta activity concentrations with this equation.

19.3 Results and Discussion

The year-to-year variation of airborne ^{210}Pb is presented in Fig. 19.2. The ^{210}Pb activity concentration in northern Finland is usually lower than that in southern Finland. Compared to southern Finland, northern Finland is more often dominated by air masses coming from the North Atlantic Ocean and the Arctic Ocean containing practically no radon, and thus no ^{210}Pb sources.

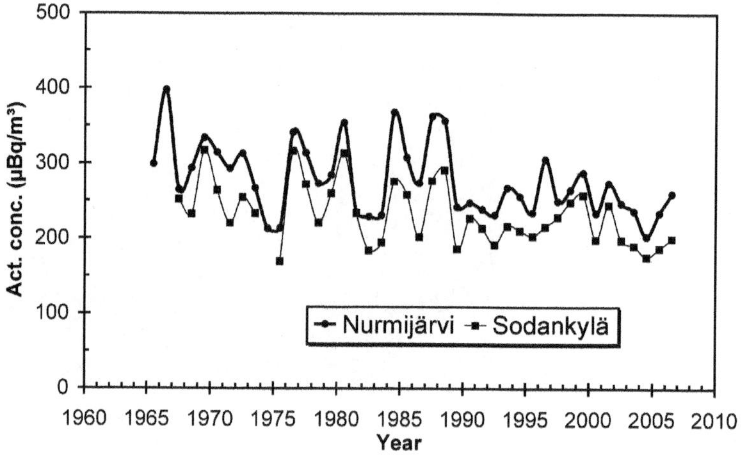

Fig. 19.2 Year-to-year variations of airborne ^{210}Pb ($\mu Bq/m^3$) at Sodankylä, northern Finland and at Nurmijärvi, southern Finland

In the past, year-to-year variations in ^{210}Pb activity concentration were explained by the North Atlantic Oscillation (NAO), in other words by heat and salinity conditions in the northern seas (Paatero et al. 1998, 2000). Despite these variations, long-term concentration levels had been constant without any evident trends. However, there seems to be a sudden decrease in these levels in the 1980s and 1990s. Starting in the late 1980s, ^{210}Pb activity concentrations in the air began to decrease (Fig. 19.3). This downward trend became steeper towards the south. Thus, in Sofia on the 42nd latitude, the downward trend is an order of magnitude stronger than at Sodankylä on the 67th latitude. Similar behavior has been reported in Germany (Winkler and Rosner 2000).

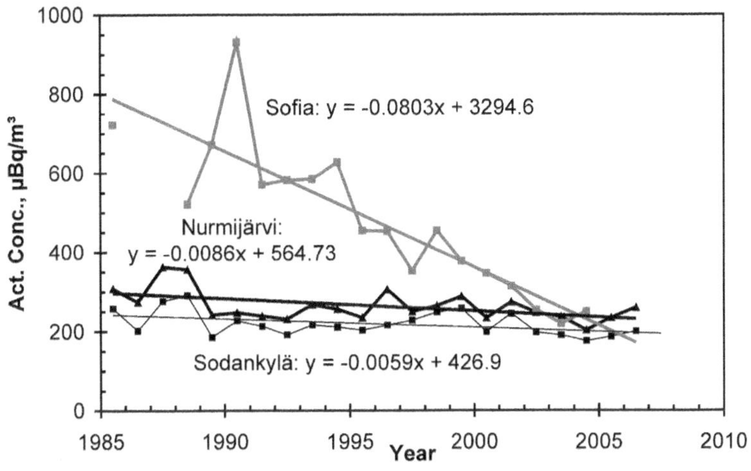

Fig. 19.3 Long-term trends of airborne ^{210}Pb (μBq/m^3) at Sodankylä, northern Finland, at Nurmijärvi, southern Finland, and in Sofia, Bulgaria

19.4 Conclusions

Especially in Bulgaria, the decrease in average activity concentrations of ^{210}Pb is quite dramatic. In less than a decade, concentrations have been reduced to a half, and the earlier-existing threefold difference between Finland and Bulgaria has disappeared. The source term of ^{210}Pb, exhalation of radon-222 from surface soil to the atmosphere, could hardly have changed that much. Therefore, either the large scale movement of air masses has changed considerably or the residence time of aerosol particles carrying ^{210}Pb in the atmosphere has shortened. The first possibility means there have been changes in the origin or type of prevailing air masses – whether continental or maritime. The second relates to precipitation amounts because aerosol particles are efficiently removed from the atmosphere through wet deposition mechanisms, both through below-cloud scavenging and especially through-cloud rainout phenomena. Our future work aims to elucidate these questions by studying long-term deposition behavior of ^{210}Pb from the northeast of Europe southwards.

Acknowledgments Cooperation between NIMH and FMI has been managed through a bilateral agreement between the Bulgarian Academy of Sciences and the Academy of Finland.

References

Baskaran M, Coleman CH, Santschi PH (1993) Atmospheric depositional fluxes of ^{7}Be and ^{210}Pb at Galveston and College Station, Texas. J Geophys Res 98:20555–20571.

Hötzl H, Winkler R (1987) Activity concentrations of ^{226}Ra, ^{228}Ra, ^{210}Pb, ^{40}K and ^{7}Be and their temporal variations in surface air. J Environ Radioact 5:445–458.

Manolov L, Teneva M (1964) Bulletin of the Institute of Hydrology and Meteorology, Vol. 1, Publishing House of the Bulgarian Academy of Sciences, Sofia.

Mattsson R, Jaakkola T (1979) An analysis of Helsinki air 1962 to 1977 based on trace metals and radionuclides. Geophysica 16:1–42.

Mattsson R, Paatero J, Hatakka J (1996) Automatic alpha/beta analyser for air filter samples – absolute determination of radon progeny by pseudo-coincidence techniques. Radiat Prot Dosim 63:133–139.

Paatero J, Hatakka J (2002) Observations of long-lived airborne and deposited radioactivity in Finland 1995–1997. Publications on Air Quality No. 30. Finnish Meteorological Institute, Helsinki.

Paatero J, Hatakka J, Mattsson R, Aaltonen V, Viisanen Y (2000) Long-term variations of lead-210 concentrations in ground-level air in Finland: Effects of the North-Atlantic oscillation. In: Transport and Chemical Transformation in the Troposphere: Proceedings of the EUROTRAC-2 Symposium 2000, Springer-Verlag, Berlin.

Paatero J, Hatakka J, Mattsson R, Viisanen Y (1998) Analysis of daily lead-210 air concentrations in Finland, 1967–1996. Radiat Prot Dosim 77:191–198.

Paatero J, Veleva B, Hatakka J, Buyukay M (2007) Airborne total beta activity in Finland and Bulgaria 1961–2000. In: International Conference on Environmental Radioactivity: From Measurements and Assessments to Regulation, 23–27 April 2007, Vienna, Austria. Book of extended synopses. IAEA-CN-145. International Atomic Energy Agency, Vienna, pp. 267–268.

Winkler R, Rosner G (2000) Seasonal and long-term variation of ^{210}Pb concentration in air, atmospheric deposition rate and total deposition velocity in south Germany. Sci Total Environ 263:57–68.

Yli-Tuomi T, Venditte L, Hopke PK, Basunia MS, Landsberger S, Viisanen Y, Paatero J (2003) Composition of the Finnish Arctic aerosol: Collection and analysis of historic filter samples. Atmos Environ 37:2355–2364.

Part V
From Global to Local and Vice Versa

Chapter 20
Drought Management Centre for South Eastern Europe

Gregor Gregorič and Andreja Sušnik

Abstract Drought is a natural phenomenon which closely connects climate and society. When drought occurs, there is certain risk that the population will suffer social and economic consequences. This risk and depth of such consequences depends on the natural frequency and severity of the drought. As the climate changes, natural hazards are increasing, and it would be reasonable to imply that social and economic risks are consequently increasing. However, natural hazards are not the only element determining risk. The other factor is society's capability to overcome difficulties caused by water shortages – i.e. vulnerability. Vulnerability determines the risk of drought impact now and in the future. Risk may rise independent of climatic trends, due to increased water demands caused by population or economic growth, or both. And the other way around: natural hazard trends may be neutralized by reducing a society's vulnerability. Assessment of both – natural hazards and societal vulnerability – are among the core objectives of the drought management centre for South Eastern Europe. Historical assessment of drought occurrence and establishment of drought monitoring systems are undertaken to establish a method for regional estimation of climatological and actual natural hazards connected to occurrence of drought. Some aspects of vulnerability (mainly in the agriculture sector) have been described for some South East European countries.

Keywords Drought · Natural hazard · Risk assessment

20.1 Introduction

No natural resource past or present has had such a profound material effect upon life as water. It represents the difference between plenty and famine, between life and death, for multitudes since the beginning of time. The availability of fresh

G. Gregorič (✉)
Environmental Agency of Slovenia, Ljubljana, Slovenia
e-mail: gregor.gregoric@gov.si

V. Alexandrov et al. (eds.), *Global Environmental Change: Challenges to Science and Society in Southeastern Europe*, DOI 10.1007/978-90-481-8695-2_20,

water defines weather and climate and is a major reason for migration of peoples across the Earth searching for better living conditions. South Eastern Europe (Albania, Bosnia and Herzegovina, Bulgaria, Croatia, Former Yugoslav Republic of Macedonia, Greece, Hungary, Montenegro, Moldova, Romania, Serbia, Slovenia, and Turkey) is not as severely affected as some parts in the world. However, drought occurrence delineates areas with insufficient water resources in this sub-region as well. There is a long history of droughts in the region and these events will continue to occur in the future, possibly with increasing impact, because of foreseen climate change and increasing climate variability.

What coping mechanisms are there for drought, land degradation and desertification? These consist of improving existent and establishing new monitoring systems for early drought detection and then applying sustainable agricultural practices ensuring food security, practicing sustainable forestry management, and stopping biodiversity loss. To achieve these goals, awareness and preparedness are required: these can be strengthened through training and technology transfer and knowledge base increases, in particular for regions prone to drought and desertification. Since drought is a regional phenomenon with diverse and complex impacts at the local level, mitigation and response actions must also be locally based in order to be appropriate. Such strategies are the main focus of the Drought Management Center for South Eastern Europe, the mission of which is to coordinate and facilitate the development, assessment, and application of drought risk management tools and policies in SEE with the goal of improving drought preparedness and reducing drought impact.

20.2 Drought Monitoring and Early Warning

It is difficult to define drought. It depends on regional differences, needs and disciplinary perspective. The absence of a single definition of drought makes it difficult to determine its onset and duration. It is not possible to monitor development of drought using a single meteorological parameter. The most frequently implemented solution is to combine various parameters into drought indicators which enable us to follow drought and its developing characteristics (Svoboda et al. 2002).

There are several indices that measure the deviation of precipitation for a given period of time from historically established norms. Some indices use a richer set of parameters than precipitation alone. For example, the Palmer drought severity index (PDSI) has been widely used in the US to determine emergency drought situations. Although a new, self-calibrating version of the index has been developed (Wells et al. 2004), we were advised while preparing the DMCSEE plan to use a newer index, the Standardized Precipitation Index (SPI; McKee et al 1993), to monitor moisture supply conditions. The SPI can be computed on various time scales and is capable of identifying emerging droughts months earlier than the PDSI. The PDSI has a rather long and fixed time constant, which is around 8 months (Szalai et al. 2000).

SPI maps are now available through the DMCSEE web page www.dmcsee.org. Maps are prepared using the widely accepted scale; drought situations are assumed to exist at SPI values below –1, with extreme drought considered to be SPI values below –2. Data from the Global Precipitation Climatology Centre (gpcc.dwd.de) is applied (Rudolf and Schneider 2005), and maps are prepared on a monthly basis. Figure 20.1 shows an example of such a map.

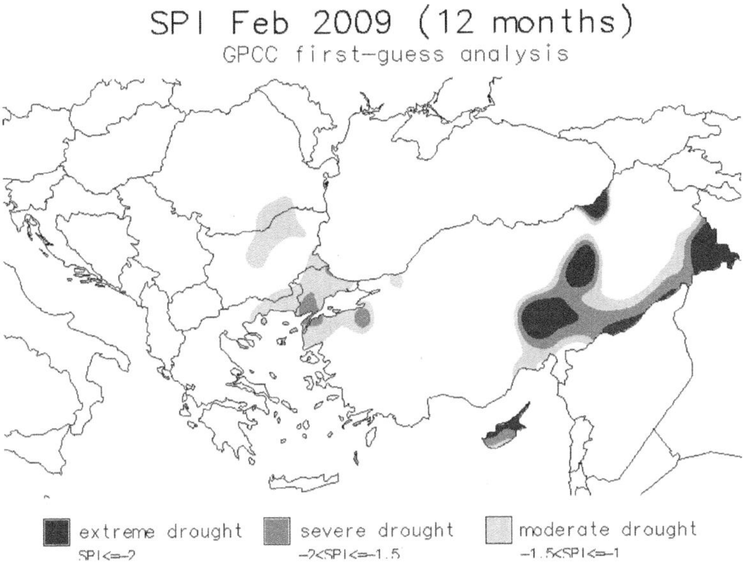

Fig. 20.1 Example of SPI map, extracted from DMCSEE web page (www.dmcsee.org)

The next task is to connect a drought index (such as SPI) to certain drought impacts. Bussay et al. (1998) and Szalai and Szinell (2000) assessed the utility of the SPI in describing drought in Hungary. They concluded that the SPI was suitable for quantifying most types of drought events. Stream flow was best described by SPIs within a time scale of 2–6 months. Strong relationships to ground water levels were found within time scales of 5–24 months. Agricultural drought (presented according to soil moisture content) was best replicated by the SPI on a 2–3 month scale. The time scale for hydrological drought was confirmed through analysis of mean monthly river discharge in central Slovenia (not shown). Although it seems logical that the first step for DMCSEE establishment is (technical) development of index-based monitoring systems, it is critically important to: (i) in parallel, study the connection between drought impact and drought monitoring indications and (ii) use available local data to ensure that appropriate details are recorded by monitoring equipment. Both aspects will be considered in the DMCSEE establishment project.

The question remains whether drought monitoring is also sufficient for the purposes of early warning. Due to slow evolution ("creeping") of drought, it seems to be possible to detect signals of emerging drought months ahead through monitoring (ground measurements and remote sensing) – for example, by comparing

SPI calculations on various time scales. However, there is a desire to look beyond that time frame, for example to assess the situation for the upcoming season. The seasonal forecasting systems that run operationally in major meteorological centres around the world seem to be able to – at least theoretically – fulfil this requirement. However, areas in mid latitudes (especially Europe) seem to have a weak signal and questionable performance, which should be examined before direct application.

20.3 Vulnerability to Drought

Vulnerability to drought is – besides the natural hazard of climate anomaly – a determination of the risk of drought impact. Vulnerability gives us the rate of response to a natural anomaly, and is equally important in determining disaster risk.

Several attempts have been made to assess vulnerability to drought. The approach of estimating the vulnerability of farm lands using GIS and remote sensing tools (Wilhelmi and Wilhite 2002) was applied in Hungary's Somogy county (Bella et al. 2005). In this case, several data layers – terrain elevation, land use, soil type, normal precipitation and groundwater levels – were combined and properly weighted. The results are presented in Fig. 20.2.

The vulnerability analysis of Somogy county was targeted at farm lands. Farm lands that are at higher elevations, more exposed to wind and solar radiation, with low groundwater levels, are marked as vulnerable. As the authors pointed out, the method is fairly simple and not too data-demanding, so it could be used in other regions and countries.

Following the Somogy example, we are preparing a map of agricultural drought vulnerability, which will synthesize a variety of data and serve as an indicator for areas deserving a detailed drought vulnerability and risk evaluation. This could aid regional decision makers in identifying appropriate mitigation actions before the next drought event.

20.4 Conclusions

Development of generally applicable methodology for drought monitoring and early warning and estimation of drought vulnerability and risks are core activities promoted by the DMCSEE. There are many possibilities for mitigating the impact of drought in the agricultural sector. The agricultural sector should therefore be the focus of DMCSEE activities. Development of monitoring systems should be followed by user-driven delivery systems and efficient decision support tools, which would enable decision makers to mitigate drought impacts.

We believe that successful execution of initial projects and sustainable functioning of the centre in the future (including intensive knowledge transfer through organization of thematic workshops and other training events) might help affected parties in the region cope with the increasing hazard of water scarcity and drought.

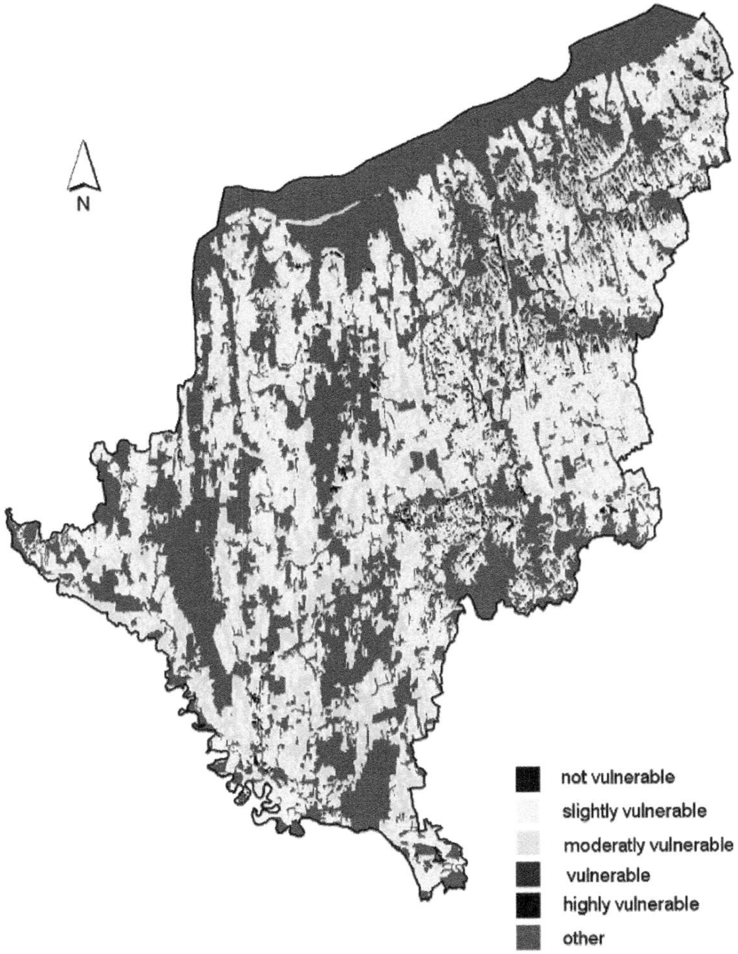

Fig. 20.2 Example of drought vulnerability analysis – Somogy county in Hungary (extracted from Bella et al. 2005)

References

Bella S, Nemeth A, Szalai S (2005) Examination of drought vulnerability using GIS tools: Somogy county case study. Goettlinger Geographische Abhandlungen 113:209–217.

Bussay AM, Szinell C, Hayes M, Svoboda M (1998) Monitoring drought in Hungary using the standardized precipitation index. The Abstract Book of 23rd EGS General Assembly, C450. April 1998, Nice; France.

McKee T, Doesken NJ, Kleist J (1993) The relationship of drought frequency and duration to time scales. Preprints, 8th Conference on Applied Climatology, pp. 179–184.

Rudolf B, Schneider U (2005) Calculation of gridded precipitation data for the global land-surface using in-situ gauge observations. Proceedings of the 2nd workshop of the International Precipitation Working Group IPWG, Monterey October 2004.

Szalai S, Szinell C (2000) Comparison of two drought indices for drought monitoring in Hungary – a case study. In: Vogt JV, Somma F (eds) Drought and Drought Mitigation in Europe, Kluwer, Dordrecht, p. 161–166.

Szalai S, Szinell C, Zoboki J (2000) Drought monitoring in Hungary. In: Early Warning Systems for Drought Preparedness and Drought Management, WMO, Geneva.

Svoboda M, LeComte D, Hayes M et al (2002) The drought monitor. Bull Am Meteorol Soc 83:1181–1192.

Wells N, Goddard S, Hayes JM (2004) A self-calibrating palmer drought severity index. J Clim 17:2335–2351.

Wilhelmi O, Wilhite DA (2002) Assessing vulnerability to agricultural drought: A Nebraska case study. Nat Hazards 25:37–58.

Chapter 21
The Role of Local Communities in Revising National Action Programs to Mitigate Drought, and Prevent and Combat Desertification: Lessons from Romania

Doru Leonard Irimie, Viorel Blujdea and Ciprian Pahonţu

Abstract An important task of signatory parties to the United Nations Convention to Combat Desertification (UNCCD) declared "affected countries" is to adopt a National Action Programme (NAP). After limited implementation of the 2000 NAP, and while confronted with the severe drought of 2007, a decision was made to proceed with its revision. A decision making process based on stakeholder participation assumes the consequent reduction of enforcement costs and enhanced effectiveness of the political programme. Another assumption is that UNCCD has an essential social component, as the decrease in agricultural production represents a great concern for rural communities in drylands. Questionnaires meant to grasp the position of local communities on drought and desertification were therefore instrumental in NAP revision. After being completed by a significant and representative number of mayoralties throughout the country, they were analysed at a centralized level. An important result of this approach was that it gave local communities key information, so they could have a thorough understanding of the processes. In addition, they seem committed to mitigation measures, by means of consistent contributions in kind and money. The exercise of NAP revision revealed several shortcomings which may be extrapolated for policy making on cross-cutting issues. First, the realization was made that the political profile of NAP relates closely to the severity of the natural phenomenon and its impact. Second, bringing together a multitude of potential stakeholders incurs transaction costs, adding to those already required for technical expertise. Last, actions proposed within an NAP may overlap with those of related programmes, such as climate change mitigation/adaptation and water management, entailing a need for synergetic action.

Keywords United Nations Convention to Combat Desertification (UNCCD) · National Action Programme (NAP) · Local communities · Questionnaires · ROMANIA

D.L. Irimie (✉)
Ministry of Agriculture, Forests and Rural Development, Bucharest, Romania
e-mail: doru.irimie@madr.ro

V. Alexandrov et al. (eds.), *Global Environmental Change: Challenges to Science and Society in Southeastern Europe*, DOI 10.1007/978-90-481-8695-2_21,

List of Abbreviations

CRIC	Committee for the Review of the Implementation of the Convention
NAP	National Action Programme
UNCCD	United Nations Convention to Combat Desertification

21.1 Introduction

One of the most important tasks of signatory parties to the United Nations Convention to Combat Desertification (UNCCD 1996) which have declared themselves "affected countries", is to adopt, and when necessary to revise, National Action Programmes to combat desertification (NAP). The Romanian Government, through the former Ministry of Waters, Forests and Environmental Protection, first adopted an NAP in 2000 and reported on the progress of its implementation at the fifth session of the Committee for the Review of the Implementation of the Convention (CRIC), in March 2007. In the absence of a legally-binding instrument to specifically address drought, the UNCCD became, in Romania, as in other affected countries, the general framework for political action.

Seven years after adoption of the previous NAP in 2000, it can be ascertained that implementation of the programme was limited for several reasons. First, public institutions are limited in their capacity to draft, promote and implement policies and strategies for sustainable development. Second, some actions and measures proposed in the NAP, such as the rehabilitation and extension of irrigation systems, were not based on a thorough analysis, such as a cost-benefit analysis, and had thus did not take into consideration the three pillars of sustainability (economic, ecologic and social), so that their application was constrained a priori. Third, some actions and measures lacked budget provisions; when funding was stipulated, the necessary financing was not allocated at the required level. In addition, responsibilities concerning the implementation and evaluation of programmes were often unclear. Fourth, the changes undergone by Romania over the last years have radically altered institutional and legal frameworks for implementation; this was an additional rationale for the amendment decision. Furthermore, the 2000 NAP had a limited grasp on inter-sectoral, cross-cutting issues, and was not promoted and supported by the weight of a governmental decision (or a document of superior rank), but only through a ministerial order. This entailed limited impact and means for implementation. Last but not least, actions and measures were characterised by centrism, so that they did not always meet the specific needs and aspirations of local communities. Local communities are supposed to be the main beneficiaries of the programme, but were only minimally involved in its drafting and implementation. These are the main shortcomings which were addressed during revision work carried out in 2007, and the present paper concentrates on the latter.

There are additional rationales behind the political decision to proceed with the NAP's revision. There is a cyclic nature to the policy making process, which starts with agenda setting, continues with policy elaboration, adoption, and implementation, and ends with evaluation (and then a new cycle). Moreover, the drought of summer 2007 was acknowledged as one of the most serious in the recorded history of meteorology, giving drought and desertification a high profile in public debate and thus pushing revision of the NAP, which became a political necessity. The Ministry of Agriculture and Rural Development, as the main organism responsible for the implementation of UNCCD, took the lead in this work.

A new trend in the decision making process, challenging the traditional assumption of self-regulating political deliberations, is to involve those primarily affected by respective policies in a framework through a bottom-up consultative process. When the aspirations and needs of people concerned are properly considered, enforcement costs will be reduced and the effectiveness of a political programme enhanced. Incentives for compliance may therefore be more efficient than penalties for non-compliance (Krott 2001).

Another assumption of NAP revision work is that UNCCD has an essential social component, which is even more evident when compared to two Rio sister conventions or to other multilateral environmental agreements (European Commission 2006). Thus, people are generally aware of various environmental issues, but they become most concerned when their livelihood is seriously threatened. While issues such as climatic change and the loss of biodiversity are very technical in nature, the decrease in agricultural production and subsequent disruptions in socioeconomic activity as a direct result of drought and land degradation cannot go unrecognised by local communities. This makes the UNCCD and subsequent processes unique among other environmental bodies, and this represented an additional impulse for undertaking the extensive work of consulting local communities.

21.2 Scope and Methodology

The main goal of the revised National Action Programme for mitigating drought, and preventing and combating desertification was to indicate actions necessary to undertake in the short, medium and long term in order to reduce the vulnerability of local communities, natural ecosystems and socioeconomic activities and minimise the combined effects of related phenomena. It represents a substantial effort in implementation of the UNCCD, as does the 10-year strategic plan and framework to enhance implementation of the Convention (2008–2018).

Based on the need to combine a traditional top-down, sectoral approach in decision making with a bottom-up one, data required for the revision of NAP were collected through three different channels: (1) eight thematic ad-hoc working groups, established for sectors of socioeconomic activity; (2) territorial departments for agriculture and rural development; and (3) local communities (communes). The first NAP draft resulting from the integration and harmonization of these data was put under public scrutiny. Following the incorporation of observations received, it was adopted by the National Committee to Combat Drought, Land Degradation and Desertification.

When designing the NAP, three main principles were followed. First, attempts were made to increase the efficiency of utilization of existing funding (e.g., state and local budgets, special funds, subsidies, etc.), rather than provisioning new funds, in order not to add an additional burden to public finances. Second, actions and measures were harmonised, as much as possible, with eligibility criteria for funding within relevant programmes of financial support, such as EU structural funds. Third, the allocation of funding on the basis of concrete projects was set as a rule, elaborated in accordance with regulations applicable for each sector and field of activity.

In Table 21.1 the framework structure of NAP is highlighted, designed on the basis of collected data. The first level of presentation on actions and measures proposed focussed on the urgency of intervention, the second on administrative/hierarchical levels of implementation, and the third on sectors (e.g., land reclamation, agriculture, forestry, etc.) and fields of activities (e.g., legislation, capacity building, monitoring/research, etc.). For each action/measure, the period or, as the case may be, the deadline for implementation was indicated, as well as the

Table 21.1 Framework structure of the programme of actions and measures to mitigate drought, and prevent and combat desertification

No	Actions/ measures	Period/ deadline for application	Responsible in implementation	Financing (amount, sources)	Responsible in evaluation, indicators	Legal basis and synergies
Actions and measures to initiate in short term						
At local level						
1	By sectors					
2						
At regional level						
At national level						
At international level						
Actions and measures to initiate in medium term						
…						
Actions and measures to initiate in long term						
…						

organisations responsible for implementation and evaluation (based on indicators of performance), financing sources, legal basis and likely synergies with relevant strategies and programmes (i.e., columns in the table). The actions/measures were also described in relation to their type (i.e., action/reaction, prevention/mitigation).

Based on assumptions concerning the role of local communities in decision making, questionnaires were elaborated, distributed and collected from representatives of local communities (i.e., mayoralties, local councils) to get an understanding of the needs, expectations and involvement of local communities in drought and desertification-related issues. A template questionnaire was elaborated both deductively (based on theory) and inductively (based on practical knowledge) and sought to uncover both quantitative and qualitative data, as can be seen in Table 21.2, which

Table 21.2 Processing of questionnaires distributed to local councils (mayoralties), concerning drought, land degradation and desertification (N = 509)

I. Reference data: completed on individual questionnaires

II. Data concerning the role of drought in the local economy

1. Land area by use categories: completed in individual questionnaires

2. Are there data/information in the administration of your locality concerning:

a.	Drought	352
b.	Land degradation (e.g.,erosion, desertification, etc.)	228
c.	Other natural hazards (e.g.,floods, landslides, tornados, etc.)	143
d.	Harmful impact of human activity (e.g.,poor cultivation practices)	94

What is the extent of the most frequent phenomenon, by land use category? (three most frequent answers given)

a.	Drought affecting grasslands	70%
b.	Drought affecting arable land	65%
c.	Drought affecting orchards and vineyards	60%

3. Land degradation occurs in the following (three most frequent) use categories:

a.	Grasslands, moderate intensity	155
b.	Arable land, moderate intensity	141
c.	Forests, none	92

Table 21.2 (continued)

4. Does drought represents a major risk for the locality?

–	Are there land areas endowed for irrigations? Yes 120 323	No
–	What is the proportion of land on which irrigation systems work?	30%
–	What is the proportion of land actually irrigated annually?	20%
–	Is irrigation equipment functional? Yes 52 No 68	

5. If relevant, what reasons describe irrigated areas which fall below those affected by drought?

a.	Lack or degradation of irrigation systems	70
b.	Associations of irrigators not established	9
c.	Water for irrigation not provided	4

6. What are the means for provision of water, by source and socioeconomic activity (three most important utilisations)?

a.	Agriculture – precipitation	80%
b.	Livestock breeding – individual underground sources	60%
c.	Population – individual underground sources	50%

7. How it is the degree of water supply for domestic consumption reated?

a. Good 293	b. Satisfactory 192	c. Not satisfactory 75

shows questionnaire results (the last section of the paper). Having been completed by a significant and representative number of mayoralties throughout the country (509 communes evenly distributed in seven development regions, representing ca. 20% of the total national number), the questionnaires were analysed at the ministry level. The presentation of questionnaire results and their evaluation in relation to the final output (NAP) are the main objectives of the next section. A detailed presentation of the whole NAP would be of interest too, but it falls outside the limited scope of the present paper.

21.3 Results

The main results summarizing questionnaire outputs are presented in Table 21.2, presented in the last section of the paper. This section concentrates on findings of

Table 21.2 (continued)

III. Data concerning socioeconomic activity

8. What is the main activity assuring revenue for the population and local budgets?

a.	Agriculture and livestock breeding, households	65%
b.	Agriculture and livestock breeding, local budgets	55%

(10% difference may be due to subsistence agriculture)

9. What are the (five) most important activities that may assure revenue in drought conditions?

a.	Commerce/trading	177
b.	Agriculture (including vegetables, orchards, etc.)	97
c.	Livestock breeding	94
d.	Micro-industry and handicrafts	84
e.	Services/tourism	82

10. If the enumerated activities above are affected by drought and other natural hazards, what are the means of improving them or alternatives to that end?

a.	Irrigation (development, reparations, new technologies)	136
b.	Management of watersheds and water sources	27
c.	Promotion of drought-resistant cultures	17

11. What are the negative impacts of drought, besides land degradation/desertification?

a.	Impinges on economic activity (including agricultural production)	339
b.	Affects wellness and working productivity, creates discomfort	376
c.	Leads to pollution; dust and other particles	187
d.	Decreases attractiveness for tourists and investors	130

12. If relevant, how are drought-related losses covered?

a.	Compensation by the state	213
b.	Utilisation of reserves and decrease in domestic consumption	156
c.	Grants on insurance contracts against natural hazard risks	152
d.	Adaptation to drought and identification of new resources	141

Table 21.2 (continued)

IV. Institutional and organisational framework at local level

13. Are there associations of land owners, agricultural producers, water users, etc.?

a.	Agricultural producers	156
b.	Livestock producers	62
c.	Water users for irrigation purposes	18
d.	Forest land owners	18

14. If not, why were they not established (first three most important)?

a.	Lack of irrigation systems	57
b.	Lack of trust in associations and cooperatives	42
c.	Lack of funding	30

15. Are there problems with their functioning, and if so, what are they?

a.	Lack of funding and access to credit	55
b.	Lack of irrigation systems	25
c.	Poor market conditions	9

16. Is there a specialist within the mayoralty to advice citizens on problems related to drought and land degradation, and if so, what does he/she do?

a.	Conveys warnings regarding natural hazards	190
b.	In forms citizens about ways and means to mitigate impacts	128
c.	Organises meetings with specialised agencies	125

particular significance for design of the revised NAP, that is, those translated into actions and measures proposed therein. The presentation of these results is made on a point-by-point basis, in accordance with the original structure of the questionnaire. A final discussion of these results takes place in the concluding section of the paper.

21.3.1 Data Concerning the Role of Drought in the Local Economy

The centralised data of the questionnaires (pct. 2 of Table 21.2) revealed that the most frequent natural hazard throughout the country is, by far, drought (352

Table 21.2 (continued)

17. Are consultancy agencies assisting you, and if so, which ones?

a.	Territorial agencies for agriculture and rural development	273
b.	Local agencies for consultancy in agriculture	159
c.	National Agency for Direct Payments in Agriculture	60

18. Are actions and measures to combat drought and desertification integrated into local development plans, and if so, what do they refer to?

a.	Management of watersheds and water sources	41
b.	Afforestation of degraded lands	35
c.	Protective forestry shelterbelts	6
d.	Irrigation networks	6

19. Are there related externally financed projects, and if so, what do they refer to?

a.	Water supply	31
b.	Afforestation of degraded lands	7
c.	Forestry shelterbelts	3

20. Do you encounter problems when drafting financing for projects, and if so, what are they?

a.	Lack of resources for co-financing	60
b.	Lack of experts/consultancy services	53
c.	Poor information	13
d.	Bureaucracy and cumbersome procedure	13

respondents, representing ca. 70% of the total). According to local knowledge perception, which confirms central statistics, the most affected land use categories are grasslands, followed by arable lands and then by orchards and vineyards. A similar classification applies in relation to the link between the processes of land degradation and land use categories in general (pct. 3). That forests are not part of this ranking confirms the high resilience of this land use category to drought and desertification.

Essential data concerning the present status of irrigation systems were also revealed by the questionnaires. Thus, of respondents addressing the respective

Table 21.2 (continued)

V. Programmes, actions and measures to mitigate drought effects

21. The National Action Programme to prevent and combat drought, land degradation and desertification should aim, by priority, at:

a.	Local development by improving the economic environment	335
b.	Rehabilitation and improvement of environmental conditions	328
c.	Measures for social protection/welfare	259

22. For a better understanding of drought and related effects, there is a need, at the local level, of additional information concerning:

a.	Monitoring of drought and its effects	366
b.	Monitoring of the status of natural resources (soil, water, vegetation)	356
c.	Research and technological development	220

23. Which legal instruments of development policy are deemed to be efficient and applicable?

a.	Imposition, through development plans, of land use categories capable of producing maximum benefit in drought conditions	316
b.	Adoption and implementation of norms for using natural resources	283
c.	Expropriation of land for public works,with compensation	135

questions (pct. 4–5), only one-third mentioned that the administrated agricultural area is endowed with irrigation systems. Furthermore, these irrigation systems are fully functional on only ca. 30% of these areas, and the area actually irrigated is smaller still. Consistent with information held at the national level, the main impediment to irrigation is a lack of, or the degradation of irrigations systems, although factors such as scarcity of water supply, and water users' associations and their service-payment capacity cannot be overlooked.

Pct. 6 of the questionnaire shows that the majority of rural populations rely on individual underground water sources for domestic consumption, including for livestock. This can be explained by the limited development of water supply infrastructure in rural areas. In spite of this situation, the greatest part of respondents (95%) consider the water supply to be satisfactory or good (pct. 7).

Table 21.2 (continued)

24. Which economic instruments of development policy are deemed to be efficient and applicable?

a.	Compensation for drought-related damages	303
b.	Co-financing(e.g., state-commune) of projects in the field	287
c.	Compensation and incentives for changing land use category or technology, or economic activities there of	248
d.	Development of market-driven instruments (e.g., certification) to reward actions and measures taken against drought	151
e.	Incentives for the consolidation of land property	144

25. Which informational instruments of development policy are deemed to be efficient?

a.	Consultancy services	382
b.	Regular information	291

21.3.2 Data Concerning Socioeconomic Activity

The questionnaires' data confirm the reliance of rural communities on agriculture and livestock breeding activities, which make up more than a half of household and local budgets (pct. 8). These activities are considered to be revenue insurance even in drought conditions, with only commerce/trading being considered more resilient (pct. 9). A large number of respondents still attach a high value to irrigation systems (136 of those who answered pct. 10), which is consistent with related national policy. It was good to see that local communities also hold valuable knowledge on "dry farming"; this was emphasized by several respondents, who raised the need to promote drought-resistant agricultural cultures.

Pct. 11 shows that most of the population acknowledge the various impacts of drought. Pct. 12 reveals, however, that they rely heavily on state compensation for agricultural losses resulting from drought, compared to insurance policies and alternative sources of revenue. These findings led to the need to incorporate measures into the NAP such as the institutionalisation of risk insurance for natural hazards.

21.3.3 Institutional and Organisational Framework at the Local Level

The questionnaires confirmed the limited development of associations at the local level (pct. 13). The apparent good development of associations for agricultural

Table 21.2 (continued)

26. What are the most appropriate measures to mitigate the effects of drought, and prevent and combat land degradation?

a.	Afforestation of degraded lands	338
b.	Rehabilitation of grasslands	327
c.	Irrigation	314
d.	Management of watersheds and water sources (e.g., wells)	305
e.	Change structure of cultures /chose new varieties	267
f.	Specialised technologies for soil cultivation	255
g.	Development of early warning systems	209

27. For which of the above-mentioned measures would local budget contribute financially, in percent of total costs?

(a) 15%	(b) and (d) 20%	(c) 10%

28. Additional proposals and suggestions of measures to be included in the strategy:

- Subventions for irrigation, land cultivation, purchasing of equipment
- Establishment of specialised structures to deal with drought and desertification
- Access to eligible UE structural funds
- Market intervention on the to assure a minimum price for agricultural products
- Development of an insurance scheme for agricultural cultures
- Increase of land area designated for extensive cultivation (e.g., forests, pastures)
- Control harvesting of forests
- Storage and use of water from precipitation

producers (156 positive answers in total) can be explained by respondents' inclination to also include, within "associations", agricultural companies established over the last years by private entrepreneurs. Limited development was otherwise explained (pct. 14–15) by a lack of irrigation systems, entailing a decrease in the production capacity of land, lack of funding (credit), poor market conditions for agricultural products, and a lack of trust in "associations", due to previous experience with communist cooperatives.

Respondents acknowledged the important role of advice agencies subordinated to the Ministry of Agriculture and Rural Development in fighting drought and desertification. Still, communities rely on centralised or decentralised agencies overwhelmingly in post-event emergency situations in order to cash the state-aid, and only rarely for continuous advice or support during normal weather conditions. For example, more than half of respondents mentioned the territorial departments for agriculture and rural development (pct. 17) among these agencies. This led to the conclusion that there is a serious need for institutional reform of decentralised territorial agencies and for enhancement of capacities to provide needed technical advice in drought situations. Such reform would also be consistent with the general policy of decentralisation of management and administration of natural resources.

The streamlining of actions and measures to mitigate drought and desertification, as well as their integration into local development plans, a critical issue in UNCCD-related debates, was assessed as generally low (pct. 18–19). Most of these measures refer to the management of watersheds and water sources, afforestation of degraded lands, establishment of forestry shelterbelts and rehabilitation of irrigation systems. The main impediments are a lack of local resources for (co)financing such projects and a scarcity of experts/consultants to elaborate and implement viable projects. These are key issues which were properly considered in the NAP revision.

21.3.4 Programmes, Actions and Measures Proposed to Mitigate the Effects of Drought

The last part of the questionnaires was designated to collect concrete proposals from local communities on content of the new NAP. The processed questionnaire data received gave a positive sign that local communities are actually capable of thinking and acting, to a large extent, in a sustainable manner, even with an ageing population, high unemployment alternatives to subsistence agriculture, spiralling costs of agriculture, and so forth. Some 2/3 of respondents acknowledged the need for improvement of the economic environment for viable programmes and local development projects, compared to approximately ½ who merely seek measures of social protection.

They also largely agreed on the need for adoption and implementation of a combined set of the three main instruments of development policy; legal (regulative), informational (monitoring, advice) and economic (incentives, subsidies, compensation). For example, more than 1/4 of local community representatives agreed that expropriation of land for public purposes, with suitable compensation, is acceptable (pct. 23, c), and more than 3/4 requested consistent consultancy and advice services (pct. 25, a).

Pct. 26–27 provides key findings regarding the level of social acceptance of main categories of actions and measures proposed through the revised NAP, translated into on-the-ground projects. The afforestation of degraded agricultural lands, rehabilitation of degraded grasslands, rehabilitation of irrigation systems, management of watersheds and wells, change in culture structures and specialised technologies for soil cultivation are concrete measures assessed on the whole as appropriate by more than half of respondents. Moreover, local communities expressed willingness and the capacity to contribute financially – up to 10–20% of total costs – for the elaboration and implementation of such projects.

21.4 Conclusions and Discussion

An important conclusion from the centralised analysis of questionnaires is that local people and their leaders hold key information for a thorough understanding of drought, land degradation and desertification issues, including aggravating factors and mitigation potential. In addition, they seem committed to mitigating drought effects, and preventing and combating desertification through consistent contributions in kind and money. A critical issue that may hinder collective action is reduced social cohesion and community spirit, diluted by decades of communism and the subsequent transformation process. Findings from the questionnaires are important not only because they emphasise the need to involve local communities, as the main beneficiaries of drought and desertification-related projects, in policy making, but also because they make us rethink the real significance and dimensions of UNCCD.

As the title suggests, the "United Nations Convention to Combat Desertification in Those Countries Experiencing Serious Drought and/or Desertification, Particularly in Africa" was designed to address the needs of people living at the extreme margin of agricultural cultivation and subsistence, where a large part of Africans exist. In relevant political circles it is often referred to as "Africa's Convention" or "the Convention of the poor". The present findings challenge this assumption to some extent, in that they show that local communities from regions considered traditionally not as badly affected by drought and desertification as the African continent also suffer from these phenomena, do hold essential traditional knowledge regarding mitigation potential, and do request active involvement in political processes. Considering the obvious social dimension of UNCCD referred to before, we could consider renaming it "the peoples" convention' and thinking of it as a real "global convention". The recently adopted 10-year strategic plan for the UNCCD and its current quasi-universal membership status support this claim.

Another important finding in the questionnaires was that while some actions and measures proposed by NAP imply a reasonable financial commitment which can be borne by local level institutions and financing instruments, many others require consistent financing which can only be supported by state and EU budgets. For example, projects on water sources management, rehabilitation of pastures or afforestation of degraded agricultural lands can be initiated and carried out based on local imitative, with limited support from central or regional authorities. However, the rehabilitation of centralised irrigation systems, research, monitoring, or establishment of a desired national network of forestry shelterbelts are examples of measures requiring consistent financing policies from central state administration.

Activities concerning the collection, centralisation and the analysis of questionnaires cannot be detached from the broader activity of NAP revision, for which it was in fact designated. NAP revision revealed several shortcomings which are often ascertained as problematic per se for inter-sectoral policy making. First, it was realised that the political will and social impact of a revised NAP relates directly to the duration and intensity of the natural hazard in question (i.e., drought). If the extent and impact of this natural phenomenon decreases under what can be considered the critical threshold, the opportunity for establishing an up-dated "programme" decreases dramatically. Second, bringing together a multitude of national authorities and organisations presumably concerned with drought and desertification, as well as the hundreds of communities where sampling was done, implies transaction costs which increase those required by technical expertise for mere elaboration of the NAP. NAP revision must rely on the contribution and the involvement of all relevant actors, but costs incurred for cooperation and participation in the process have to be considered from the beginning. Finally, actions and measures proposed in this strategy may overlap with those proposed under similar initiatives (e.g., adaptation to and mitigation of climate change, water management, etc.), so that the critical issue of synergizing actions comes again to the fore. It can safely be maintained here that synergies between the numerous multilateral environmental agreements will be a standing issue in the relevant political agenda for years to come.

The revised NAP is no invention whatsoever. It may indeed be considered a step forward in comparison to what has been elaborated so far, yet its intrinsic value and applicability should not be overestimated. The NAP is inter-sectoral and does not aim at solving the numerous existing sectoral problems, even though actions and measures were divided by fields and sectors of socioeconomic activities for administrative purposes. Furthermore, it relies heavily on knowledge, communication and cooperation between different organisations. Its success therefore greatly depends on the capacity of people running these organisations to understand drought and desertification issues as a result of both unfavourable natural conditions and poor management practices, mobilise resources for implementation, provide necessary feedback and cooperate with all interested actors, that is, to change the "business as usual" practice.

Acknowledgments The authors are grateful to all those involved in the demanding work of distribution and completion of the questionnaires: the representatives of local communities (mayoralties) and the officers within territorial departments for agriculture and rural development in Romania.

References

European Commission, Directorate General JRC (2006) Technical paper to provide input to the Inter-sessional Intergovernmental Working Group (IIWG) of the United Nations Convention to Combat Desertification: Review and reform of the scientific and technical functions of the UNCCD.

Krott M (2001) Strategies of the state forest service in European wide comparison 1991–2000. In: Krott M (ed) EFI Proceedings 40: 41–64.

UNCCD (2009) The ten-year strategic plan and framework to enhance the implementation of the Convention (2008–2018), Decision 3/COP8. In: Report of the Conference of the Parties on its Eight Session, held in Madrid from 3 to 14 September 2007. URL: http://www.unccd.int/cop/officialdocs/cop8/pdf/16add1eng.pdf

United Nations (1996) United Nations Convention to Combat Desertification in Those Countries Experiencing Serious Drought and/or Desertification, Particularly in Africa. URL: http://www.unccd.int/convention/text/convention.php

Chapter 22
Strengthening Capacities in Western Balkan Countries to Address Environmental Problems Through Remediation of High Priority Hot Spots

Stewart Williams and Sladan Maslać

Abstract UNDP country offices in Albania, Bosnia and Herzegovina, Macedonia FYR, Montenegro, Serbia and the UN-administered Province of Kosovo have developed a regional demonstration programme around demand-driven projects in nine locations in the Western Balkans suffering from the legacy of polluting industries and requiring industrial renewal, environmental cleanup and new economic initiative. The approach of the 30-month, approximately 15 million USD programme is to achieve improvement of the environmental situation and quality of life for citizens living in and around polluted areas through least-cost measures, improved local and national policy dialogue and supply of domestic professional services in the environmental management sector. While the main focus will be on physical works required to mitigate ecological problems, institutional strengthening and capacity-building will be important subjects throughout the programme.

Keywords Western Balkans · Environmental programme · Hot spots · Remediation · Clean-up · Capacity building

22.1 Introduction and Context

Environmental issues have not been among the top national priorities in the Western Balkans. Understandably, priorities to date have focused on the reforms needed to strengthen security, rebuild the economy and to improve general living conditions. As a result, much- needed investment in environmental infrastructure such as wastewater treatment, air-pollution abatement and monitoring, and industrial and

S. Williams (✉)
Western Balkans Environmental Programme, United Nation Development Programme Montenegro, Podgorica, Montenegro
e-mail: stewart.williams@undp.org

V. Alexandrov et al. (eds.), *Global Environmental Change: Challenges to Science and Society in Southeastern Europe*, DOI 10.1007/978-90-481-8695-2_22,

communal waste management are still waiting their turn. Clearly, this is a situation that raises humanitarian, social, economic and environmental concerns.

In addition, a number of industrial towns and regions face a complex challenge related to past industrial development and pollution legacies. The environmental situation in these hot spots is a direct cause of poor health and related poverty and presents a major barrier to future investments and related economic opportunities for the local population. On one hand they face requirements for environmental clean-up and on the other they are struggling with the problems of poverty, lack of infrastructure and services and lack of prospects for younger generations.

Because of the proximity and geographic connectedness of the Balkans, the ongoing or potential pollution from these hot spots has a significant cross-border impact on air and water quality (rivers, lakes and the sea). The situation is often further complicated by uncoordinated or even conflicting plans and demands of various sectors of government, business and society. Success in cleaning up these locations and addressing environmental problems will attract new investments from inhabitants themselves, private sector, banks, donors etc.

Against this background, UNDP country offices in Albania, Bosnia and Herzegovina, Macedonia FYR, Montenegro, Serbia and the UN-administered Province of Kosovo have developed a regional programme around demand-driven projects in selected locations in the Western Balkans suffering from the legacy of polluting industries and requiring industrial renewal, and environmental cleanup. The approach of the 30-month, approximately 15 million USD programme is to achieve improvements in the environmental situation and quality of life for citizens living in and around polluted areas through least-cost measures, improved local and national policy dialogue and a supply of domestic professional services in the environmental sector.

22.2 Intervention Strategy

The Programme "Strengthening capacities in the Western Balkan countries to address environmental problems through remediation of high priority hot spots" is focused on environmental cleanup and remediation of polluted industrial hot spots that have already been identified as high priority but have seen limited progress so far due to complexity of issues and the costs involved. The programme is contributing to achievement of the following overall objectives in the region:

(A) Increased capacity of national and local governments in the Western Balkan to implement sustainable environmental policies in accordance with EU standards;
(B) Strengthened regional cooperation in the Western Balkan to solve problems of cross-border contamination due to industrial and mining activities.

Concrete project objectives are:

(i) Secure progress in clean-up of nine priority physical sites, raise awareness and strengthen capacity for good environmental management at local and national levels;
(ii) Enhanced regional cooperation through improved information-sharing progress and professional consultancy services regarding good environmental management.

The planned targets for the programme are:

(1) Planned clean-up works in nine hot spot locations completed, in accordance with technical specifications, thus contributing to a reduction of existing/potential pollution;
(2) Memorandum of Understanding/Cost Sharing Agreements for hot spots signed by municipal authorities and local stakeholder groups;
(3) Assessment Reports with follow-up measures prepared;
(4) 70% of the total value of the clean-up contract awarded to local companies or institutions;
(5) Two workshops/trainings/seminars per year organized;
(6) TV clip prepared and at least three newspaper articles published on the project;
(7) Three pilot projects on policy integration implemented, promoting sustainable environmental management;
(8) Three training courses/seminars/workshops per year organized;
(9) Market survey of domestic experts from business or NGO sector conducted;
(10) Database of qualified local experts established;
(11) One regional high-level/ministerial meeting per year organized promoting cooperation and information sharing between governments in the region;
(12) One study tour per year organized for stakeholders in the region;
(13) All programme activities effectively coordinated and implemented;
(14) Organization of the programme itself in three components.

22.2.1 Component 1: Clean-Up Projects

This is be the main component of the programme and is formed around environmental remediation work in high priority hot spots in the region. The hot spot sites were selected during the programme preparation phase from among those already identified by the ENVSEC initiative or national authorities through National Environmental Action Plans as top priority hot spots. Local authorities and/or site owners at the potential project sites were invited to submit information detailing:

(a) Description of the current situation including environmental, social and economic aspects;
(b) Summary of past activities (if any) and their evaluation;
(c) Indicative activities under this programme with cost estimates;
(d) Secured or potential matching funds;
(e) Willingness of the authority and other stakeholder to actively participate in the programme and disseminate its results.

Based on the interest expressed (in form of letters of commitment), particular hot spots were selected in each participating country/territory through consultation between the UNDP, relevant national authorities, local Netherlands's Embassies, and civil society. The criteria used for selection were as follows:

(1) Trans-boundary effects of hotspots;
(2) The possibility to attract other donor funding;
(3) Possibility to establish partnerships, including public-private partnerships;
(4) Readiness of local authorities and relevant stakeholders to take leadership in sustainable environmental management of their community.

Five of the nine initial pilot or demonstration projects (no's 4, 5, 6, 8 and 9) were *mining hotspots*, most of them also identified as highest priority clean-up candidates by the EnvSec[1] (Initiative) desk study and proposal on environmental security risk reduction from mining.

One of these five (Mojkovac/Montenegro) concerns *abandoned* mining operations, and four (Bucim and Lojane/Macedonia and 2 × Trepca/Kosovo) address *re-opened* mining and minerals processing activities.

Two projects, both in Bosnia and Herzegovina, are concerned with changing the local energy supply system for a particular user group (two health clinics) at the municipal level in order to close down two heavily polluting coal fired plants for the (public health) benefit of its own urban population.[2]

One project (the Kula-Vrbas section of the Danube-Tisa-Danube or Grand Canal/Serbia) addresses just one component of a multi-year, large investment (more than USD 50 million) water resources rehabilitation; without such prior rehabilitation a desirable level of integrated WR management can never be achieved.

One project (Shkoder/Albania) addresses the issue of a relatively simple hotspot careless storage of a mix of toxic chemicals and other materials with possibly long lasting negative effects on a very valuable (Ramsar-listed) ecological system.

[1] Environmental Security Initiative.

[2] This intervention can be compared with another on a much larger scale, the KEK energy supply in Kosovo, also being supported by the Netherlands Regional Environmental Program for the Western Balkans.

Altogether, this package is believed to represent a good "cross section" of the hotspot problems in the Western Balkan region, each calling for tailor-made corrective actions – or at least a first step in the right direction.

The establishment of a water and/or air pollution *monitoring* regime and system a fundamental responsibility of the hotspot "owner" is a common requirement for all pilot or demonstration projects, albeit one that is to be elaborated upon in a project-specific manner. In the absence of a ("minimal") water and/or air pollution *monitoring* regime and system, it would not be possible to assess the effectiveness of the concerned clean-up activity.

For each hot spot an action plan has been developed and agreed upon at the steering committee level, based on environmental assessments and technical proposals for remediation works. Project funds are provided to each site in accordance with the needs identified in the action plan, contribution of own resources and absorption capacity, ranging from 1.1 to 1.6 million USD. The targeted overall matching funds from hotspot site owners for each location is up to 50% (in actual work or cash), which constituted a criterion for selection of location in the first place.

The implementation of all nine action plan cleanup projects was launched in the first 6 months of the programme; implementation and is to be finished within the second year in order to allow for evaluation and dissemination of results. In order to achieve broader dissemination and public awareness raising in and among the countries, a documentary has been commissioned which is covering all of the hot spots, describing the initial situation, project efforts, achieved results and the commonality and value in a regional approach. These documentaries will be made available to all TV networks in the region and internationally in the final phase of the project.

Links are being established with trained journalists in the area, to encourage them to report on project activities.

22.2.2 Component 2: Demonstration and Information Sharing

This component is aimed at: (a) promoting the necessity for countries to move from legislation to implementation in protecting environment and managing natural resources[3] (b) offering "low hanging fruits", i.e. demonstrating concrete results of the benefits of good environmental management to local communities through concrete demonstration projects.

The goal is to involve (mostly through workshops/trainings/meetings/media campaigns, etc.) relevant institutions of national government for environmental management and sustainable development, including environmental ministries, but

---

[3]Although most environmental legislation is in place, implementation has been slow and therefore progress in protecting the environment and managing natural resources is not yet evident throughout the region.

also core "economic" ministries such as finance, industry and labor, public institutions (hydro-metrological and ecotoxicology institutes) as well as civil society (networks of NGOs, chambers of commerce). Specific activities are being organized with the aim of supporting resumption of regional cooperation.

The main objective is to promote (and advocate for) sectoral integration as an important element in the EU SAP/accession process. In this regard, three regional meetings on sustainable environmental management are planned to be will be organized involving national integration frameworks and relevant line ministries.

22.2.3 Component 3: Strengthen Supply of Professional Consultancy Services

In order to efficiently meet its objectives and secure sustainability and multiplication of its results, the implementation of the programme is based on the involvement of domestic/regional professionals (individuals, private sector and public institutions) in project demonstration activities at the local (hot-spot) and national (policy analysis) level where possible.

At the beginning of the project, a market survey was conducted of domestic experts from business or NGO sectors already working in the field of policy development and implementation, with particular emphasis on those with experience related to environmental management and sustainable development. CVs of potential experts have been collected and are presented in a database.

This roster will serve to efficiently identify, select and contract experts for specific assignments under the various project activities. The roster is also being made available to local/national authorities for use in their own procurement procedures as well as to other donors and the private sector.

22.2.3.1 Relationship Between Components 1, 2 and 3

The "demonstrable relevance" of each of the selected and agreed-upon projects provides an important link to *component 2: Demonstration and Information sharing*. Not only the "inputs", i.e. the rationale of the (9) individual interventions, summarized in the middle column of Table 1, are to be shared among all interested stakeholders in the region, but also what these will achieve (the *measured* effects), because this is what should convince the concerned parties (central governments/ministries of environment, ministries of energy, ministries of industry, ministries of economy, etc., autonomous provinces, municipalities, private and public sector entrepreneurs) to ultimately assume their legal responsibilities.

"Policy integration" mechanisms emphasizes Environmental Impact Assessment (EIA) as the primary instrument or tool to be applied properly and respected by all concerned parties listed above. So, *Demonstration and Information sharing* activities, both at the national and regional levels, are probably best articulated through dealing in training and workshops with all "ins and outs" of EIAs within the "thematic" range of the 9 pilot schemes.

Furthermore, Environmental Impact Assessments undertaken so far in all six countries, say since the year 2000, offer perhaps the most concrete and practical reference for taking stock of currently available regional expertise, applying a simple check list:

(a) Does the ministry (or department) of environment (protection) have a department/section dealing with the (legally required?) EIA studies?
(b) What is the level of education/training/experience specifically focused on EIA studies (approximating EU standards) of these particular staff?
(c) Is there a complete list of EIA studies of certain (to be specified) size and complexity (e.g. excluding simple "building permit" like exercises) in each of the countries since the year 2000? – a "short-list" of most relevant EIA studies per country.
(d) Is there a list of all key experts from the "six nations" region who were substantially involved (e.g. explicitly mentioned as co-authors) in these more sizeable and complex EIA studies? An experienced, independent expert must prepare a "quality assessment" of the short-listed EIA studies (package of all six countries) – a desk exercise, resulting in a (quality) ranking and a "cut-off point" of just acceptable quality;
(e) Prepare a short-list of (only) those key experts from the "six nations" region, that were substantially involved in these EIAs that passed the quality-threshold – and request the complete CVs (Resumés) from these experts in order to verify and to understand the full expertise of this group of experts.

Adopting this "logic" of relationship between components 1, 2 and 3 will also facilitate a more concrete specification of both the main tasks (job content) and the required skills and experience of the key project staff, i.e.

(1) Regional Programme Manager (RPM)
(2) Chief Technical Advisor (CTA)
(3) National Project Coordinator (NPC)

The project team is organized with the Project Implementation Unit (PIU) based in Podgorica, Montenegro and national coordinators in UNDP offices in each country/territory. This is because UNDP Montenegro is designated the UNDP subregional (Western Balkans) environmental focal point; it has available expertise and an established regional network, as well as being geographically the most appropriate coordination point in the region. The Chief Technical Advisor is part of the PIU, based in UNDP Montenegro with frequent travel in the region. The advisor provides high-level policy advice to programme and country/territory activity leaders.

The programme, throughout its regional activities, has strongly promoted the necessity of regional networking, because of the necessity to increase regional communication and exchange experience among experts and civil servants. In addition, the supply of development services and related multiplication of best practices will be increased through networking among national institutions, local communities and environmental governance practitioners in the region.

In engaging Western Balkan societies in regional activities, relating in this case to environmental hotspots, it is with the aim of sharing resources, ideas and experiences in solving common problems that are difficult to solve when societies act alone. The potential benefits of such cooperation extend beyond the hotspot issues and pave the way for further cooperation on other expected environmental challenges, including the impact of climate change including warming and droughts which are expected in the coming years.

Acknowledgments I would like to thank the UNDP Montenegro Western Balkans Environmental Programme team for its generous support in preparation of this paper.

References

UNDP Montenegro programme team, 2007. "Strengthening capacities in the Western Balkans countries to address environmental problems through remediation of high priority hot spots". *Programme document.*

ENVSEC Report: Draft of the refined assessment of the South Eastern European mining related risks: Identification and verification of "environmental hot spots" for Albania, Bosnia & Herzegovina, Macedonia, Montenegro and Serbia, 2006.

UNEP Technical Report: From Conflict to Sustainable Development, Assessment of Environmental Hot Spots, Serbia and Montenegro, April 2004.

UNEP Final Report: Strategic Environmental Policy Assessment - FYR of Macedonia; UNEP Final Report: Post-Conflict Environmental Assessment-Albania, 2000.

Chapter 23
Pilot Training Programme for Integration of Global Environmental Concerns into the Regional and Spatial Planning Process in Bulgaria

Nataliya Dimitrova-Popova

Abstract Environmental issues have traditionally been underestimated in decision making processes at national, regional and local levels in Bulgaria. Until recently, there has been no effective mechanism for their integration into regional development and spatial planning policy. The major obstacles have been a lack of experience in implementation of up-to-date procedures, methodologies and approaches for such integration, especially at lower levels of regional planning; problems with generation and processing of primary (environmental) data; and a lack of unified standards for data maintenance and exchange.

To alleviate those problems, the Ministry of Regional Development and Public Works (MRDPW), Ministry of Environment and Waters (MoEW) and the United Nations Development Program (UNDP), supported by the Global Environmental Fund (GEF), initiated a project in 2006 entitled *Integrating Global Environmental Issues into Bulgaria's Regional Development Process (Working title: The Rio Conventions Project)*. The project objective is to build MRDPW and MoEW capacities in district administrations and municipalities in Bulgaria in an effort to mainstream global environmental issues into the formulation and implementation of regional and local development, as well as spatial planning policies.

Keywords Regional development · Capacity building · Public institutions · Global environment

List of Abbreviations

CEU	Central European University
EA	Environmental Assessment
EIA	Environmental Impact Assessment
EU	European Union
GE	Global Environment

N. Dimitrova-Popova (✉)
Rio Conventions Project, United Nations Development Programme, Sofia, Bulgaria
e-mail: natalia@rioconventions.org

V. Alexandrov et al. (eds.), *Global Environmental Change: Challenges to Science and Society in Southeastern Europe*, DOI 10.1007/978-90-481-8695-2_23,

GEF	Global Environmental Facility
GIS	Geographic Information System
MoEW	Ministry of Environment and Water
MRDPW	Ministry for Regional Development and Public Works
NCSA	National Capacity Self Assessment
NGO	Non-Governmental Organization
NOPRD	National Operational Program for Regional Development
NSFRD	National Strategy for Regional Development
NTTA	Natura 2000 Assessment
OPs	Operational Programs
RCP	Rio Conventions Project
RDA	Regional Development Act
RDPs	Regional Development Plans
SA	Sustainability Appraisal
SD	Sustainable Development
SEA	Strategic Environmental Assessment
SEE	South Eastern Europe
SU	Sofia University
TP	Training Program
UD&SP	Urban Development and Spatial Planning
UNCBD	United Nations Convention on Biological Diversity
UNCCD	United Nations Convention to Combat Desertification
UNDP	United Nations Development Programme
UNEP	United Nations Environment Programme
UNFCCC	United Nations Framework Convention on Climate Change

23.1 Introduction

The GEF-funded Rio Conventions Project (RCP) (full title: *Integrating Global Environmental Issues into Bulgaria's Regional Development Process*) aims at promoting proactive integration of global environmental issues into the very process of regional and local development, as well as spatial planning, both of which are managed by the Ministry of Regional Development and Public Works (MRDPW). This could be achieved by developing the capacity of MRDPW and the Ministry of Environment and Water (MoEW) to integrate global environmental objectives into regional and local development policies and practices, as well as into spatial planning documents. The project is being implemented between November 2006 and October 2010.

A National Capacity Self Assessment[1] (NCSA) conducted in 2002–2004 and subsequent discussions with key stakeholders during the Project Development

[1] http://chm.moew.government.bg/ncsa/indexEn.htm

Phase, identified that proactive integration of global environmental issues into the very process of regional and local development in Bulgaria still does not take place. One of the reasons for this gap is lack of a targeted and in-depth training program in Bulgaria to provide sufficient theoretical background and practical skills to public officials and experts in governmental institutions on the integration of global environmental objectives into regional and local development policies and practices, as well as into spatial planning documents. Hence, one of the major objectives of the project is to develop a training program for public officials and other experts in the cross-cutting areas of environmental protection and regional development.

In July 2008, UNDP commissioned the task of creating a training program to two universities working in cooperation: the Central European University, Budapest, Hungary, and the Sofia University, Sofia, Bulgaria.

23.2 Scope and Structure of the Training Program

23.2.1 Scope of the Training Program

The Training Program Package consists of three related training courses which all cover the same general topics and which differ only in training duration: the introductory course is 1–2 days long, the "core" training course about 10 days, to be conducted over a period of 1–2 months, and a two or three semester-long academic course brings one to the Masters' level.

"Core" training is aimed at about 130 ministry staff (MRDPW, MoEW) and other experts who are working on planning, implementation and monitoring of plans and programmes related to regional development. As the trainees are all part of the ministerial workforce, it is considered prudent to hold different training sessions over a period of 1–2 months and to keep individual training sessions down to a maximum of 2 days.

A key component of the "core" training course is that the trainees should, upon completion of the course, be able to act as mentors/trainers for ministerial staff with whom they work. In other words, the quality of training should be very high, providing the trainee with sufficient information, examples and practical knowledge to enable him/her to effectively share his/her new knowledge. It is, therefore, important that the course has a *final exam* and a certification. Passing a final exam and obtaining a certificate should guarantee that trainees who have passed the program are capable of engaging in peer-to-peer training and transfer of knowledge.

The developed "core" training course has a good balance of theory, case studies and practical exercises and covers the general topics outlined below. Based on the "core" training course a "light" 1–2 day introductory course was prepared. This introductory course highlights key messages of the "core" training. The introductory course fits the needs of a particular group of trainees: for key governmental decision makers at middle and upper management levels, it stresses the importance and relevance of integrating environmental concerns into regional and spatial planning; for senior municipal planners it provides an introduction to global environmental issues in general.

The academic course is also based on the "core" training course, which principally consists of 360 academic hours of training (30 credits). It expands on practical exercises and includes more in-depth analytical assignments and submission of case papers. It was officially included in the Sofia University catalogue of Master courses in September 2009.

23.2.2 Training Program Documentation

To support the training a series of materials were developed, including a minimum of power-point presentations supporting all training sessions for all three courses listed above, as well as trainee manuals for "core" and "introductory" training courses providing additional background material. All training materials have already been published on the Sofia University web site at the following link: http://www.gis.gea.uni-sofia.bg/rioconventions/?q=node/19. The training materials also ensure that trainees have a full and in-depth understanding of individual sessions and contain an outline for further suggested reading for the "core" and "introductory" training courses, which may encourage trainees to engage in further self-motivated study. Materials for practical exercises and case studies were also developed for the "core" and "academic" training courses.

23.2.3 Structure of the Training Program

All three courses mentioned above focus on the integration of environmental protection, including bringing UNCBD, UNFCCC, and UNCCD[2] objectives into regional development and spatial planning processes, as well as planning and implementation related to EU structural funds. The training consists of the following general sections:

23.2.3.1 Background Section

This section gives the trainee general knowledge of international agreements and initiatives, including the three Rio Conventions, as well as EU legislation directives and initiatives related to the cross-cutting area of sustainable development and environmental protection. The trainee should also become familiar with how these EU and international agreements that Bulgaria is signatory to are transposed into

[2]United Nations Convention on Biodiversity, United Nations Framework Convention on Climate Change and United Nations Convention on Desertification.

the national legislature. This section is primarily based on presentations, but group discussions are also encouraged.

The background section should cover the following key topics:

- History, definition, principles and milestones of sustainable development
- The Rio Conventions, i.e.: United Nations Convention on Biological Diversity (UNCBD), Cartagena Protocol on Biosafety, United Nations Framework Convention on Climate Change (UNCCC), Kyoto Protocol; United Nations Convention to Combat Desertification (UNCCD); Regional implementation Annex for Central and Eastern Europe
- Other international conventions and treaties Bulgaria is a party to, i.e.: Vienna Convention on the Protection of the Ozone Layer, wetlands protection
- EU trends and policies regarding integrating SD and environmental protection into regional development and spatial planning
- Bulgaria's implementation of commitments and requirements for integrating sustainable development and environmental protection into regional development and spatial planning
- Review of planning processes in Bulgaria

23.2.3.2 Toolbox Section

The Toolbox section provides trainees with sufficient knowledge of the main tools commonly used in ensuring integration of environmental protection – including UNCBD, UNFCCC, and UNCCD objectives – into regional development and spatial planning processes, as well as planning and implementation related to EU structural funds. Key assessment processes such as Natura 2000, Environmental Impact Assessment (EIA) and Strategic Environmental Assessment (SEA) (in Bulgaria referred to as Ecological Assessment) are a main component of the section. Through this section, trainees should obtain a full understanding of the overall process of conducting different types of impact assessments, as well as why this process is important to ensure sustainable development. In addition, trainees gain the skills needed for developing, implementing, monitoring and reporting on national, regional and local plans, as well as strategies for regional development and spatial planning. To this end, knowledge of how to identify indicators and how to grasp commonly used (by EU, GEF etc.) indicators, is of the essence. This section is based on presentations, as well as group discussions and practical exercises aimed at furthering understanding.

The Toolbox Section should cover the following key topics:

- Processes and requirements of the "Environmental Assessment" and "Environmental Impact Assessment" and how understanding these processes and requirements can lead to better and more cost-effective planning

- Processes and requirements of the "Natura 2000 Assessment"[3] and how understanding these processes and requirements can lead to better and more cost-effective planning
- The use of indicators and why they are important
- Indicators based on technologies such as Geographic Information Systems and remote sensing systems
- The importance of ToRs (Terms of Reference) spell out in ensuring on-the-ground implementation of sustainable development and environment protection concerns

23.2.3.3 Practical Skills Section

This section allows trainees the possibility to utilize what they have learned in the previous two sections in practical and applicable training exercises and to gain confidence in using this information to conduct work-related exercises. It is also hoped that at the end of this section the trainee will have gained practical expertise to be able to communicate and act as a mentor in his/her workplace. The main practical exercise involves trainees demonstrating how they would approach the subject of incorporating sustainable development and UNCBD, UNFCCC, and UNCCD objectives into regional development and spatial planning processes.

This section should cover the following key topics:

- Sustainable development and environment protection: a topic for all Bulgarians
- Key areas to address when integrating global sustainable development concepts into regional development and spatial and urban planning
- Examples of Bulgarian strategic and legislative framework on how well sustainable development and the Rio Conventions are taken into consideration
- Incorporating sustainable development and UNCBD, UNFCCC, and UNCCD objectives into regional development and spatial planning processes
- Possible areas of concern, i.e.: water and sewage supply in residential areas, decreasing urban pollution, improving urban infrastructure and the physical urban environment, economic development areas, social development fields, management of settlements and municipalities, capacity-building and institutional development

23.2.3.4 Self-Evaluation and Testing Section

This section provides the trainee with an opportunity to review, on a daily basis throughout the training by means of specially designed evaluation forms, how he/she perceives and how he/she can use newly acquired knowledge in daily work; this

[3] Assessment of plans and projects significantly affecting Natura 2000 sites.

exercise also helps in identifying areas where more training is needed. This section consists of a final exam in which the capacity and knowledge of trainees is tested and evaluated. This testing assists in evaluation of the overall training course.

As a result of this section trainees should be able to answer the following questions:

- How do I ensure that I fully consider sustainable development and environmental protection, including the Rio conventions, in my daily work regarding (legislation, regulations, orders, rules, guidelines, instructions, methodologies, correspondence, distribution of information, coordination, decision making)?
- Do I need more training to be able to fully ensure that I can integrate sustainable development and environment protection, including the Rio conventions, into my daily work?

23.3 Follow up Conventions Project

As a result of joint efforts between the two universities (CEU and SU) a pilot "core" training course was developed in the spring of 2008 and tested in five consecutive sessions of 2 days each during April and May 2008 at Sofia University in Bulgaria. Fifteen participants from MRDPW, MoEW, district administrations and municipalities took part in the training.

On the basis of feedback received from trainees during pilot testing of the training program package, the contents and training materials of three types of courses were updated and translated into Bulgarian. Full-scale delivery of the "core" training course commenced in September 2008 at Sofia University. It is expected that by March 2010, 130 public officials from MRDPW, MoEW, municipalities and district administration will have attended the course. Delivery of the "core" course was guaranteed through signing of an MoU between UNDP, MRDPW and MoEW, which establishes the responsibilities of each organization in recruitment of participants for the course and in covering the associated costs of training and participation.

Delivery of the "introductory" course took place between January and June 2009, when the course was delivered "on-site" to 27 district administration centers in Bulgaria by Sofia University professors, and to more than 200 public officials from the above-mentioned institutions.

The developed "academic" course was officially approved by the Sofia University Academic Board in the summer of 2009 and was included in the 2009 Fall Catalogue of Master Courses. The Master Course will be taught in two and three semester options.

Index

V. Alexandrov et al. (eds.), *Global Environmental Change: Challenges to Science and Society in Southeastern Europe*, DOI 10.1007/978-90-481-8695-2,

GPSR Compliance

The European Union's (EU) General Product Safety Regulation (GPSR) is a set of rules that requires consumer products to be safe and our obligations to ensure this.

If you have any concerns about our products, you can contact us on ProductSafety@springernature.com

In case Publisher is established outside the EU, the EU authorized representative is:

Springer Nature Customer Service Center GmbH
Europaplatz 3
69115 Heidelberg, Germany

Batch number: 10761599

Printed by Printforce, the Netherlands